J. P. Jin
Q. Q. Huang

Calponin and Transgelin

Molecular Evolution, Function, Regulation and Medical Relevance

DOI: https://doi.org/10.52305/OZEL6040

Library of Congress Cataloging-in-Publication Data

ISBN: 979-8-89530-468-6 (Softcover)
ISBN: 979-8-89530-551-5 (eBook)

Published by Nova Science Publishers, Inc. † New York

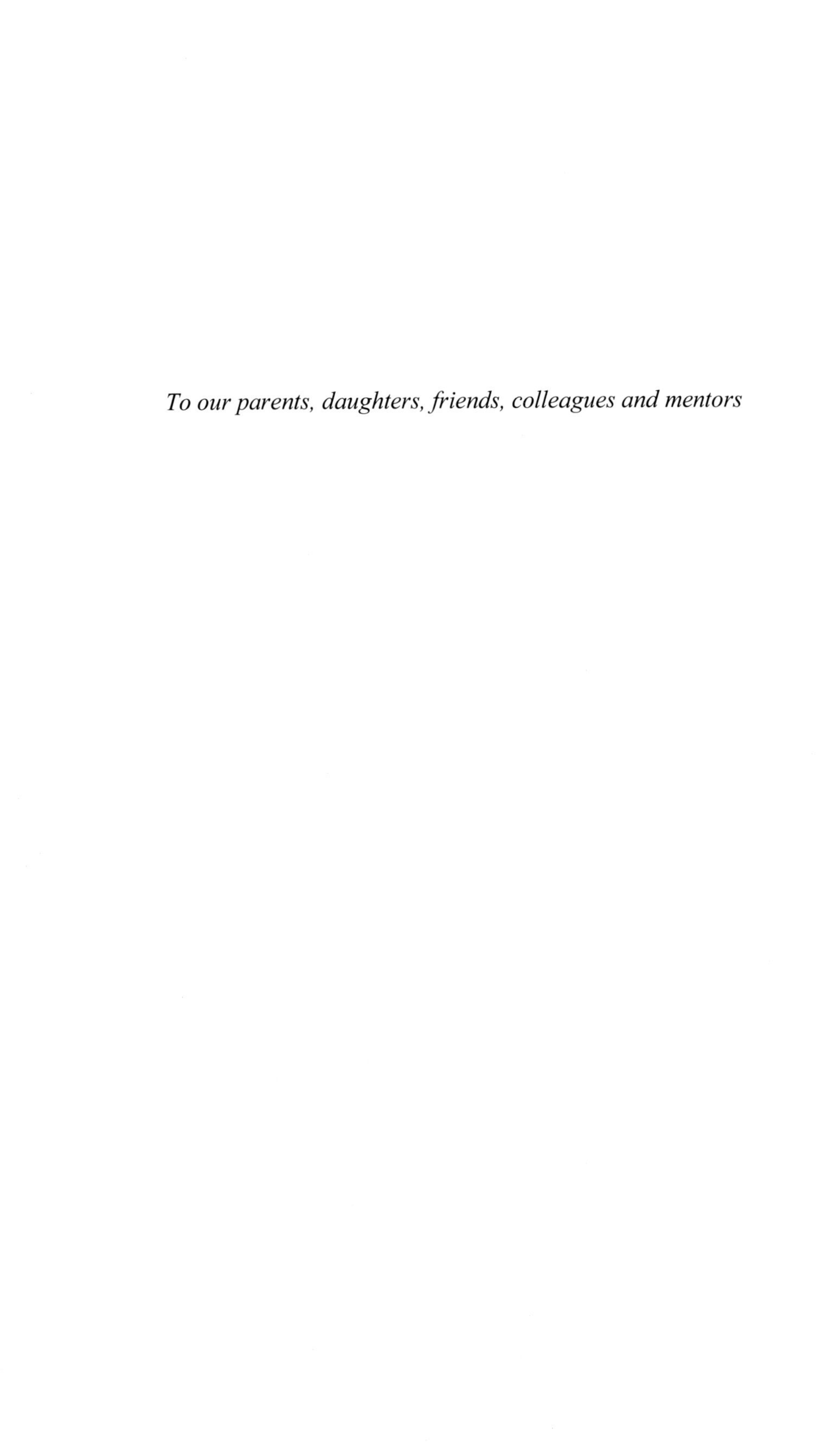

To our parents, daughters, friends, colleagues and mentors

Contents

Acknowledgments

This book would not be possible without the dedicated work of our research team over the past three decades, including the contributions of many outstanding laboratory members and collaborators. We are especially grateful for the research accomplishments made since 1993 by Drs. Jimin Gao, Rita Nigam, M. Moazzem Hossain, Han-Zhong Feng, Wenrui Jiang, Airong Qian, Patrick Hines, Rong Liu, Olesya Plazyo, Tzu-Bou Hsieh, and Monica Rasmussen.

Introduction: The Calponin-Transgelin Family of Actomyosin ATPase Regulatory Proteins

Calponin was first found in chicken gizzard smooth muscle as a striated muscle troponin T-like protein that binds the actin filament with a proposed function in regulating smooth muscle contraction (Takahashi, Abe, et al., 1988; Takahashi et al., 1986). Extensive research followed and demonstrated that calponin is a family of actin filament-associated regulatory proteins of 34-37 kDa (292-330 amino acids) in size and expressed in both smooth muscle and non-muscle cells to function as an inhibitory regulator of actin-activated myosin ATPase and motor activities. Three isoforms of calponin (calponin 1, 2 and 3) are found in vertebrates encoded by three homologous genes (*CNN1*, *CNN2* and *CNN3*) (R. Liu & Jin, 2016a).

In vitro protein binding studies demonstrated that calponin binds actin (Takahashi et al., 1986; Winder & Walsh, 1990) and cross-links actin filaments (B. Leinweber et al., 1999). Calponin also binds and interacts with many other cytoskeleton-related proteins and regulatory molecules, including Ca^{2+}-calmodulin (Takahashi et al., 1986), tropomyosin (Childs et al., 1992; Takahashi, Abe, et al., 1988), myosin (Szymanski & Tao, 1997), caldesmon (Graceffa et al., 1996), α-actinin (North et al., 1994), desmin (Mabuchi et al., 1997; P. Wang & Gusev, 1996), tubulin (Fujii & Koizumi, 1999), gelsolin (Ferjani et al., 2006), Ca^{2+}-S100 (Fujii et al., 1994), and phospholipids (Bogatcheva & Gusev, 1995). Functional significances of these biochemically detected bindings of calponin to cytoskeleton proteins and regulatory molecules have been a focus of active research. With the localization to actin filaments (Czurylo et al., 1997; B. Leinweber et al., 1999), calponin modulates the functions of smooth muscle myofilaments and contractility. It also regulates the actin cytoskeleton of non-muscle cells in various motility-related cellular activities, such as cytokinesis, phagocytosis, cell adhesion, migration and fusion.

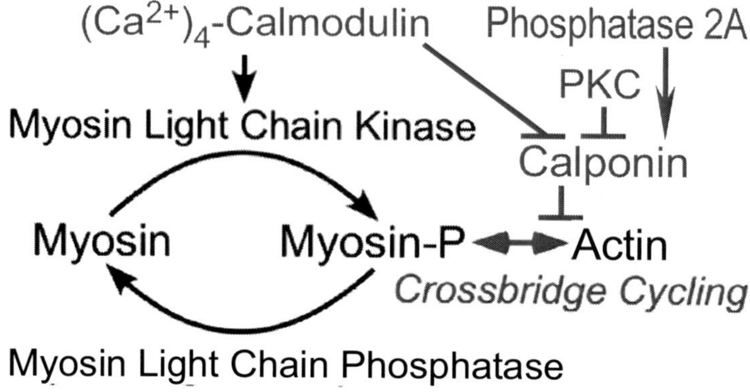

Figure 1. The primary function of calponin as an inhibitory regulator of actomyosin ATPase. While the myosin ATPase and motor activities in smooth muscle and non-muscle cells are primarily regulated by the phosphorylation of myosin regulatory light chain, calponin binds to the actin filament and inhibits actomyosin ATPase in a parallel thin filament-based regulation like that of troponin in striated muscles. The calponin inhibition is regulatorily released by the binding of Ca^{2+}-calmodulin and/or protein kinase C (PKC) phosphorylation of calponin which is reversable by protein phosphatase 2A (Walsh, 1991).

Transgelins are a group of cytoskeleton proteins homologous to calponin (T.-B. Hsieh & Jin, 2023). Transgelin was originally identified as a 22-kDa protein in chicken gizzard smooth muscle and named SM22 (Lees-Miller, Heeley, & Smillie, 1987; Pearlstone et al., 1987). Transgelin/SM22 is one of the most abundant proteins in vertebrate smooth muscles with a function in stabilizing actin cytoskeleton, similar to that of calponin (Gimona et al., 2003). Encoded by three homologous genes (*TAGLN*, *TAGLN2* and *TAGLN3*), three isoforms of transgelin, transgelin-1 (SM22α), transgelin-2 (SM22β) and transgelin-3 (SM22γ), are present in vertebrate animals (T. B. Hsieh & Jin, 2023; J. Liu et al., 2020).

Homologs of calponin and transgelin family proteins have also been found in invertebrates, such as unc-87 in *Caenorhabditis elegans* (Goetinck & Waterston, 1994) and mp20 in *Drosophila melanogaster* (Ayme-Southgate et al., 1989). Calponin is present in invertebrate muscles which are structurally similar to that of vertebrate smooth muscles (Royuela et al., 1997).

Nearly four decades after the discovery of calponin and SM22, the physiological functions of the calponin-transgelin family proteins remain to be fully established. Based on the fundamental activity of calponin as an inhibitory regulator of actin-activated myosin ATPase (Figure 1), a comprehensive collection of the current knowledge on the molecular evolution, tissue and cell type-specific expression, structural and functional features, gene regulation and posttranslational modifications of calponin-transgelin family proteins can help researchers of various background and interests to further explore their biological functions and structure-function relationships with physiological and pathological insights. The goal of this book is to review and discuss representative research data with an emphasis on the roles of this family of cytoskeleton regulatory proteins in cell motility-related processes and mechanoregulation to stimulate continued investigations.

Non-Standard Abbreviations

A/C ratio	albumin/creatinine ratio
AVIC	aortic valve interstitial cell
BMP	bone morphogenetic protein
BSA	bovine serum albumin
CAVD	calcific aortic valve disease
CH domain	calponin homology domain
CLIK	calponin-like
CTF	cell traction force
ER	endoplasmic reticulum
ERK	extra-cellular regulated kinase
FBS	fetal bovine serum
G/B ratio	glomerular tuft to Bowman's capsule area ratio
GBM	glomerular basement membrane
GPI	glucose-6-phosphate isomerase
HES-1	hairy and enhancer of split 1
IQR	median ± interquartile range
KO	knockout
LDL	low-density lipoprotein
mAb	monoclonal antibody
MAP	mean arterial pressure
MAPK	mitogen-activated protein kinase
MEF	mouse embryonic fibroblasts
MES	2-[*N* morpholino]ethanesulfonic acid
MHC	myosin heavy chain
MOF	multi-oocyte follicle
NCM	net contractile moment
NMR	nuclear magnetic resonance
OCP	osteoclast precursor
Otx2	orthodenticle homeobox 2
PAGE	polyacrylamide gel electrophoresis

PBS	phosphate buffered saline
PC3-M	metastatic derivative of prostate cancer cell line PC3
PFP	primary foot process
pI	isoelectric point
PKC	protein kinase C
POI	premature ovarian insufficiency
PRR	pattern recognition receptor
RBP	recombining binding protein suppressor of hairless
RMST	root mean square traction
ROCK	Rho-associated kinase
Runx2	Runt-related transcription factor 2
SDS-PAGE	SDS-polyacrylamide gel electrophoresis
SMA	smooth muscle actin
TBS	Tris-buffered saline
TRAP	tartrate-resistant acid phosphatase
TSE	total strain energy
WT	wild type.

Chapter 1

Molecular Evolution of Vertebrate Calponin Isoforms

Natural selection drives genetic diversity as well as conservation. Molecular evolution data provide powerful information to guide experimental investigations to understand the functional significance of the regulation of gene expression and protein structure-function relationships. This approach is especially valuable for understanding the physiological functions of calponin and its isoforms.

1.1. Three Calponin Isoform Genes in Vertebrates

Three isoforms of calponin encoded by three homologous genes are present in vertebrates: A basic calponin (calponin 1, isoelectric point, pI, = 9.4) encoded by the *CNN1* gene (J. Gao et al., 1996), a neutral calponin (calponin 2, pI = 7.5) encoded by the *CNN2* gene (Masuda et al., 1996; Strasser et al., 1993), and an acidic calponin (calponin 3, pI = 5.2) encoded by the *CNN3* gene (Applegate et al., 1994). In the human genome, *CNN1* is on chromosome 19p13.2 (Miano et al., 1997), *CNN2* is on 19p13.3 (OMIM entry 602373), and *CNN3* is on 1p22-p21 (Maguchi et al., 1995).

A nomenclature of "*h*-calponin" was used in the literature (the three calponin isoforms were then named h1-, h2-, and h3-calponins) based on the molecular weight of calponin 1 in comparison to a low molecular weight "*l*-calponin" that was seen in human urogenital smooth muscle protein extract (Draeger et al., 1991). However, such low molecular weight isoform or splice form of calponin has not been confirmed by any evidence of mRNA sequences, genomic DNA organizations, or mass spectrometry protein sequences. Therefore, the "*l*-calponin" band observed might be a product of proteolysis. To avoid confusion, the current literature uses calponin 1, calponin 2, and calponin 3, respectively, for the protein products of *CNN1*, *CNN2* and *CNN3* genes.

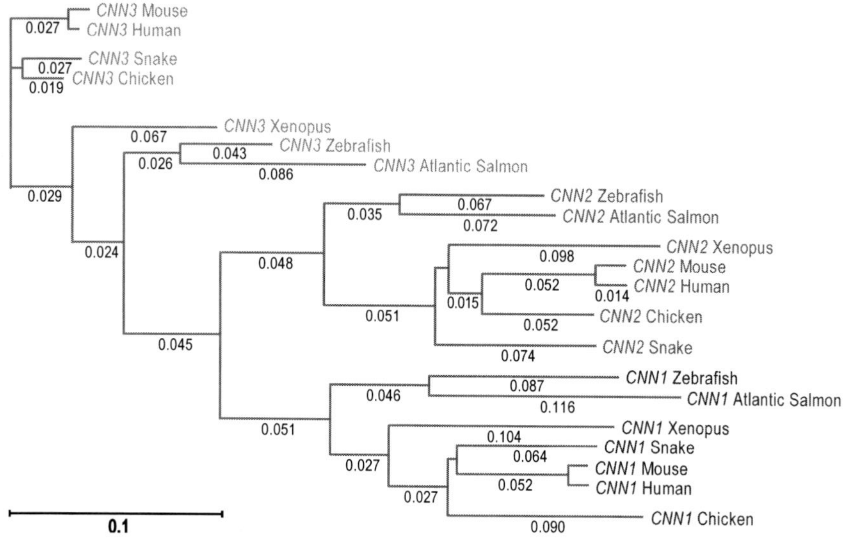

Figure 2. Evolutionary lineage of vertebrate calponin isoforms. The phylogenetic tree is generated by alignment of amino acid sequences of calponin isoforms in representative vertebrate species with the DNASTAR MegAlign computer software (Lasergene, lnc, Madison, WI) using the Clustal W method. The sequence similarity-derived evolutionary lineage demonstrates that each of the calponin isoforms is conserved in the vertebrate phylum while the three isoforms have significantly diverged during vertebrate evolution. This pattern indicates adaptations of the calponin isoforms to different cellular functions and tissue environments with differentiated functions. The degrees of evolutionary divergence are indicated by the lengths of lineage lines. The NCBI database accession numbers of the sequences analyzed are: Atlantic salmon *CNN1*, NP_001139857.1; Atlantic salmon *CNN2*, NP_001133873.1; Atlantic salmon *CNN3*, NP_001133337.1; Zebrafish *CNN1*, XP_701038.5; Zebrafish *CNN2*, NP_998514.1; Zebrafish *CNN3*, NP_956047.1; African clawed frog *CNN1*, NP_001085014.1; African clawed frog *CNN2*, ABG49504.1; African clawed frog *CNN3*, NP_001080482.1; Western terrestrial garter snake *CNN1*, XP_032066879.1; Western terrestrial garter snake *CNN2*, XP_032064388.1; Western terrestrial garter snake *CNN3*, XP_032074010.1; Chicken *CNN1*, NP_990847.1; Chicken *CNN2*, NP_001135728.1; Chicken *CNN3*, NP_001341600.1; Mouse *CNN1*, AAI38864.1; Mouse *CNN2*, EDL31614.1; Mouse *CNN3*, AAH85268.1; Human *CNN1*, NP_001290.2; Human *CNN2*, AAI48265.1; Human *CNN3*, AAB35752.1.

Comparison of mRNA (cDNA) sequences and the deduced protein primary structures demonstrated that calponins 1, 2 and 3 have largely conserved structures. The phylogenetic tree in Figure 2 constructed by alignment of amino acid sequences of the three calponin isoforms of representative species in the vertebrate phylum shows s striking pattern that each of the isoforms is conserved across vertebrate classes whereas the three isoforms have significantly diverged during evolution. The structure conservation of each of the diverged calponin isoforms reflects the adaptation of the calponin isoforms to their tissue and cellular environments with differentiated biological functions.

Gene targeting studies showed that mice with the knockout (KO) of either *CNN1* (Matthew et al., 2000; Takahashi et al., 2000) or *CNN2* (Huang et al., 2008) gene are viable and fertile. In contrast, *CNN3* KO resulted in fetal and neonatal lethality due to defective development of the central nervous system (Flemming et al., 2015) The essential role of calponin 3 in embryonic development is consistent with the more ancestral emergence of *CNN3* gene than that of *CNN1* and *CNN2* genes (Figure 2). The notion that calponin 3 represents a prototype of vertebrate calponin from which the three present-day isoforms have evolved is supported by a combined phylogenetic lineage analysis of vertebrate and invertebrate calponin genes (T.-B. Hsieh & Jin, 2023).

1.2. Tissue and Cell Type-Specific Expression of Calponin Isoform Genes

The three calponin isoform genes show different profiles of tissue and cell type-specific expressions, indicating functions corresponding to the specific cellular environment and/or physiological states.

1.2.1. Calponin 1

The expression of *CNN1* gene that encodes calponin 1 is highly specific to differentiated smooth muscle cells (Figure 3A). Most of the biochemical studies of calponin have been obtained from experiments using chicken gizzard smooth muscle calponin 1. The expression of *CNN1* gene is upregulated postnatally to reach a high level in adult smooth muscle organs (Hossain et al., 2003). It may serve as a marker of smooth muscle differentiation and maturation (Duband et al., 1993). The *CNN1* gene was activated by angiotensin II-induced G-protein signaling. ERK1/2 might be involved in this signaling pathway (Dulin et al., 2001). In vivo experimental data also showed that angiotensin II upregulates *CNN1* gene expression in rat aorta (Castoldi et al., 2001). CArG box is a representative transcriptional factor binding element in genes encoding smooth muscle contractile proteins. The activation of CArG box was linked to the contractile phenotype of vascular smooth muscle cells. *CNN1* gene contains conserved CArG-rich regions, which is necessary for the smooth muscle-specific expression of calponin 1 in vivo (Long et al., 2011).

The levels of calponin 1 vary among different types of smooth muscle. High levels of expression are seen in phasic smooth muscles of the digestive tract and only a very low level of expression was detected in avian trachea (Figure 3A) (J. P. Jin et al., 1996). The high expression of calponin 1 in differentiated contractile smooth muscles (Hossain et al., 2003) is rapidly downregulated when growth-arrested contractile smooth muscle cells re-enter the G1 phase of cell cycle to proliferate (Samaha et al., 1996). The mature smooth muscle cell-specific high-level expression of calponin 1 indicates its role in smooth muscle differentiation and contractile functions.

There have been reports of calponin 1 expression in non-muscle cells in culture dishes, such as human glomerular mesangial cells (Sugenoya et al., 2002), pancreatic precursor cells (Morioka et al., 2003), testis Sertoli cells (Zhu et al., 2004), and kidney periglomerular myofibroblasts (S. Y. Lee et al., 2010). While these observations were primarily based on detections using antibodies of which the specificity between calponin isoforms need to be confirmed, the expression and function of calponin 1 in cells cultured in vitro on the high stiffness plastic substrates are likely correlating to a myofibroblast phenotype (Masur et al., 1996).

1.2.2. Calponin 2

Calponin 2 is expressed in a broad range of tissue and cell types, including mature smooth muscles, developing and remodeling smooth muscles (Hossain et al., 2003), epidermal keratinocytes (Fukui et al., 1997; Hossain et al., 2005), lung alveolar cells (Hossain et al., 2006), endothelial cells (J. Tang et al., 2006), kidney podocytes (T. B. Hsieh & Jin, 2022), fibroblasts (Hossain et al., 2005), aortic valve interstitial cells (Plazyo et al., 2018), myoblasts (Jiang et al., 2014), myeloid blood cells (Huang et al., 2008), platelets (Hines et al., 2014), inner ear hair cells (L. Y. Zhou et al., 2023), and prostate cancer cells (Moazzem Hossain et al., 2014).

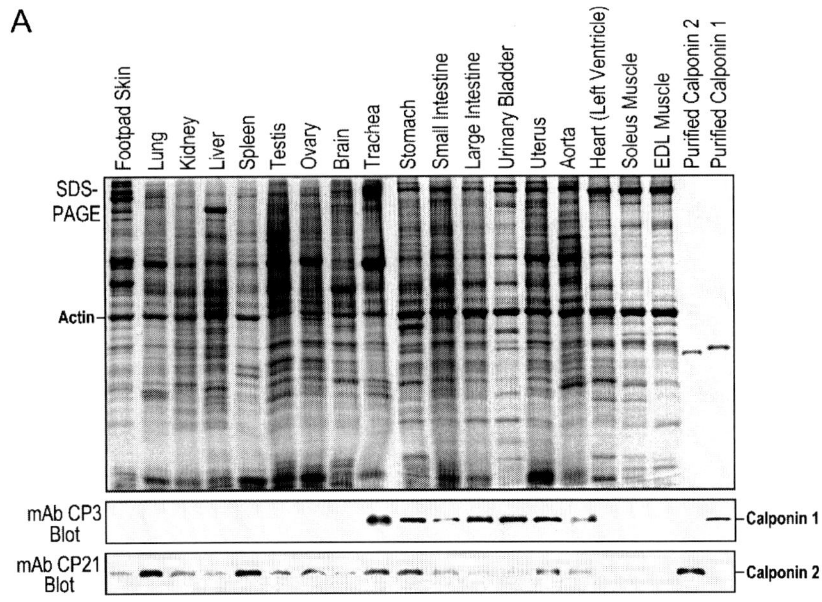

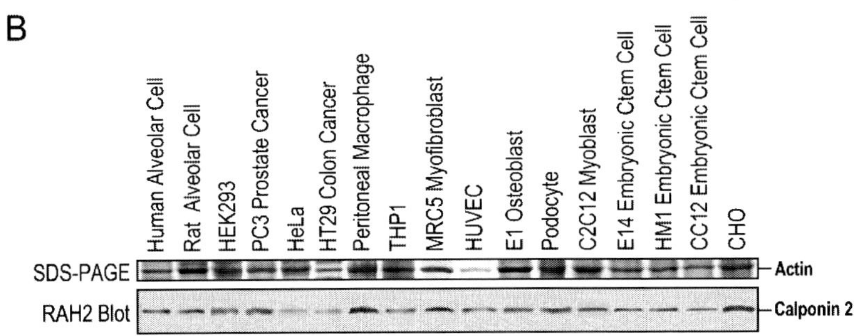

Figure 3. Expression of calponin 1 and calponin 2 in representative tissue and cell types. A. Total protein extracts from major organs of the mouse were analyzed with SDS-polyacrylamide gel electrophoresis (SDS-PAGE) and Western blots using an anti-calponin 1 monoclonal antibody (mAb) CP3 and an anti-calponin 2 mAb CP21. Purified mouse calponin 1 and calponin 2 expressed from cloned cDNA in *E. coli* cultures were used as control. The blots detected calponin 1 only in smooth muscle-rich organs at various levels relative to that of actin and calponin 2 in smooth muscle and multiple non-muscle organs with high levels of expression in the spleen and lung. B. Western blot analysis of total cellular protein extracts using a rabbit anti-calponin 2 polyclonal antibody RAH2 showed the expression of calponin 2 in multiple cell types represented by the immortalized cell lines and mouse primary peritoneal macrophage. The amounts of protein input are shown by the actin bands in the accompanying SDS-gel.

Those multiple cell types (Figure 3B) can be placed into three groups: a) cells that are physiologically under high mechanical tension, e.g., smooth muscle cells in the wall of hollow organs, epithelial and endothelial cells, lung alveola and renal glomerular podocytes; b) cells that have high rates of proliferation, e.g., myoblasts and epidermal keratinocytes; and c) cells that are actively migrating, e.g., fibroblasts and macrophages. The tissue and cell type distributions of calponin 2 reflect its role in regulating cytoskeleton tension and cell motility-based functions, stability of the actin cytoskeleton, and cell interactions with the extracellular mechanical environment (R. Liu & Jin, 2016a; Rasmussen & Jin, 2024).

1.2.3. Calponin 3

Calponin 3 is found in the brain with expressions in neurons (Represa et al., 1995; Trabelsi-Terzidis et al., 1995), astrocytes (Agassandian et al., 2000) and glial cells (Ferhat et al., 1996). It may function in regulating the actin cytoskeleton with a proposed role in the plasticity of neural tissues (Ferhat et al., 2003; Plantier et al., 1999). Calponin 3 is expressed in myoblasts and embryonic trophoblasts with a likely function in cell fusion during myogenesis and implantation of embryos (Shibukawa et al., 2010, 2013). Calponin 3 also expresses in B lymphocytes whereas the functional significance remains to be investigated (Flemming et al., 2015).

1.3. Functional Significance of Calponin Isoform Divergence

Research using genetically modified mice has elucidated the physiological significance of the isoforms of calponin. Mice with double KO of *CNN1* or *CNN2* genes survive but show additive changes in smooth muscle contractility (Feng et al., 2019). Although calponin 1 and calponin 2 show some degree of mutual exchangeability in modulating smooth muscle contractility (Feng et al., 2019), their presence did not compensate for the developmental defect caused by *CNN3* KO (Flemming et al., 2015), consistent with the importance of isoform-specific functions. On the other hand, the use of a *cytomegalovirus* promoter-driven transgene for unregulated systemic over-expression of calponin 1 in the construction of transgenic mice did not yield any live founders, implicating embryonic lethality caused by cell type-nonspecific and/or non-physiologically regulated expression of calponin (our unpublished result).

While calponin evolutionarily emerged before the divergence of vertebrates and invertebrates (T.-B. Hsieh & Jin, 2023), the three isoforms of vertebrate calponin have largely conserved structures. Therefore, their core functions are conserved and their tissue- and cell type-specific functions may be based on differentially regulated expression in different cell types and in different developmental or tissue differentiation states. On the other hand, the notable evolutionary divergence among the three calponin isoforms, such as their significantly different C-terminal structures, may fine tune their biochemical activities, targeting to different subcellular compartments and/or crosstalk with different partner molecules (R. Liu & Jin, 2016a).

More studies are needed to fully understand the biological significance of the three calponin isoforms. The molecular evolutionary lineages and structural conservation versus divergences between isoforms and among species adapted for different environmental or physiological selections can inform new investigations to better understand the structure-function relationship of calponin and the three isoforms as well as their physiological functions in various cell types and organ systems. Participating in the GeneWiki project, we have summarized the three vertebrate calponin isoform genes in the following Wikipedia links:

- https://en.wikipedia.org/wiki/Calponin_1,_basic,_smooth_muscle
- https://en.wikipedia.org/wiki/Calponin_2
- https://en.wikipedia.org/wiki/Calponin_3,_acidic.

Chapter 2

Biochemical and Cellular Functions of Calponin

The functional domains of calponin have been extensively studied. Summarized in the linear map in Figure 4, the conserved structures of the three vertebrate calponin isoforms include a calponin homology (CH) domain, two actin-binding sites, and three repeating motifs. These conserved functional structures form a foundation for the core biological function of calponin and the isoforms in regulating various actin filament-based cellular activities. The main difference among the three vertebrate calponin isoforms is the C-terminal variable segment that differs significantly in lengths and sequences.

2.1. Biochemical Function of Calponin and Phosphorylation Regulation

Unphosphorylated calponin binds to F-actin and inhibits actomyosin ATPase (see Figure 1 in the Introduction). Protein kinase C (PKC) phosphorylation of Ser_{175} of calponin 1 alters the molecular conformation and lowers the binding affinity for F-actin (J. P. Jin et al., 2000). The biochemical consequence of PKC phosphorylation is releasing the inhibition of actomyosin ATPase, which increases the contractile force of smooth muscle (Gerthoffer & Pohl, 1994; D. C. Tang et al., 1996; Winder, Allen, et al., 1993; Winder & Walsh, 1993).

Despite the overwhelming evidence for the phosphorylation regulation of calponin function from in vitro studies, phosphorylated calponin is not readily detectable in vivo or in living cells under physiological conditions (Barany & Barany, 1993; Gimona et al., 1992). A likely reason might be that since PKC phosphorylation of calponin weakens the binding affinity for F-actin (Jin et al., 2000), the phosphorylated calponin will dissociate from the actin cytoskeleton and be rapidly degraded in the cell as that shown during cytokinesis (Qian et al., 2021).

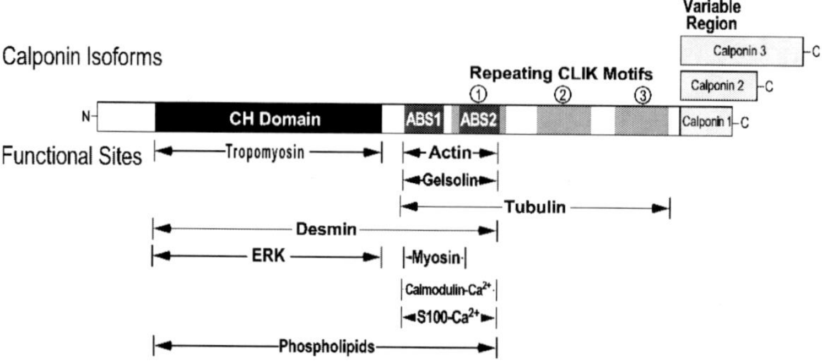

Figure 4. Structural and functional domains of vertebrate calponin. Multiple cytoskeleton and regulatory proteins have been found to interact with calponin. Summarized from numerous protein binding studies, the linear map shows that calponin binds actin filaments through two actin-binding sites (ABS), tropomyosin via the N-terminal calponin homology (CH) domain, and gelsolin via the actin binding sites. Calponin interacts with microtubules through the actin-binding sites and the calponin like (CLIK) repeating motifs and with desmin through the region of CH domain and the actin-binding sites. The segment of amino acid residues 144–182 of calponin 1 interacts with myosin. Ca^{2+}-calmodulin and Ca^{2+}-S100 bind calponin at the actin-binding sites, which releases calponin's inhibitory effect on myosin ATPase. An N-terminal fragment of calponin was reported to interact with phospholipids and the CH domain overlaps with an extracellular signal-regulated kinase (ERK) binding region. Besides the conserved structural and functional domains, the C-terminal segment of calponin is a variable region that is significantly diverged in length and amino acid sequence among the three vertebrate calponin isoforms. No protein binding site is found in the C-terminal variable region.

PKC signaling may also regulate calponin distribution in cells as calponin 1 undergoes a PKC agonist-induced translocation from the central contractile actin filaments to the submembranous cortex (Parker et al., 1994). In addition to being a contractile regulatory protein, calponin may serve as an adaptor protein connecting the PKC signaling cascade (through phosphorylation of Ser_{175}) to the ERK cascade (through the CH domain, Figure 4) to provide a scaffold for the ERK signaling complex (B. Leinweber et al., 2000).

2.2. Calponin 1 in the Regulation of Smooth Muscle Contractility

Calponin 1 was the first calponin isoform identified in differentiated smooth muscle cells (Takahashi et al., 1986; Takahashi, Hiwada, et al., 1988) and its function in regulating smooth muscle contractility has been extensively investigated. Results of numerous in vitro studies demonstrated that calponin 1 functions as an inhibitory regulator of smooth muscle contraction through inhibiting actin-myosin interaction and myosin ATPase (Allen & Walsh, 1994; Takahashi, Abe, et al., 1988; Takahashi et al., 1986). In this regulation, binding of Ca^{2+}-calmodulin or PKC phosphorylation dissociates calponin 1 from the actin filament, releases the inhibition of myosin ATPase and increases smooth muscle contractile force (Naka et al., 1990).

There are also ex vivo and in vivo experimental data supporting the role of calponin 1 as a regulator of smooth muscle contractility. Vas deferens smooth muscle from calponin 1 KO mice showed significantly faster maximum shortening velocity than wild type (WT) control, consistent with a loss in the inhibition of myosin ATPase (Takahashi et al., 2000). On the other hand, aortic smooth muscle of Wistar Kyoto rats that naturally lacks calponin 1 is contractile but shows decreased sensitivity to norepinephrine activation (Facemire et al., 2000; Nigam et al., 1998). Consistently, proteolysis of calponin 1 by matrix metalloproteinase-2 in endotoxemic rats resulted in vascular hypocontractility when stimulated with phenylephrine (Castro et al., 2012). Calponin 1 KO mice exhibit a resting mean arterial pressure same as that of WT mice but a blunted blood pressure response to phenylephrine treatment (Masuki, Takeoka, Taniguchi, & Nose, 2003). These data show that although calponin 1 is not essential for smooth muscle to contract, it functions as a regulator of smooth muscle contractility in response to physiological activations.

Calponin 1 co-localizes with and stabilizes actin filaments in smooth muscle cells. In response to the activation of PKC, it translocates from the central contractile bundles consisting of smooth muscle actin and myosin to peripheral podosomes that consists of actin and cell adhesion-related proteins (Dykes & Wright, 2007; Parker et al., 1998), suggesting that calponin 1 also regulates the non-contractile actin cytoskeleton in smooth muscle cells.

2.3. Calponin 2 in the Regulation of Actin Cytoskeleton in Non-Muscle Cells

Calponin 2 is expressed in smooth muscle and multiple types of non-muscle cells, including epidermal keratinocytes (Hossain et al., 2005), lung alveolar cells (Hossain et al., 2006), prostate gland (Moazzem Hossain et al., 2014) and kidney epithelial cells (Jiang et al., 2014), endothelial cells (J. Tang et al., 2006), kidney podocytes (T. B. Hsieh & Jin, 2022), ovarian granulosa cells (T.-B. Hsieh & Jin, 2024), fibroblasts (Hossain et al., 2005), aortic valve interstitial cells (Plazyo et al., 2018), myoblasts (Jiang et al., 2014), inner ear hair cells (L. Y. Zhou et al., 2023), myeloid blood cells (Huang et al., 2008), platelets (Hines et al., 2014) and lymphocytes (Flemming et al., 2015). Like calponin 1, calponin 2 binds F-actin and is associated with the actin cytoskeleton of smooth muscle and non-muscle cells. Transfective expression of calponin 2 in cultured cells that lack endogenous calponin increased the resistance of actin stress fibers to cytochalasin B depolymerization, indicating a role in stabilizing the actin cytoskeleton (Hossain et al., 2005).

Like calponin 1 in smooth muscle cells, calponin 2 regulates cell motility through inhibiting actin activated myosin motor functions. An in vivo study detected calponin 2 mRNA in the protrusions at the leading edge of migrating neural crest cells. Knockdown of the expression of *CNN2* gene in these cells caused randomized outgrowth of protrusions and migration defects accompanied by central stress fiber formation and reduced peripheral actin network (Ulmer et al., 2013). To demonstrate a common function of calponin 2 in inhibitory regulation of cell migration, primary fibroblasts and peritoneal macrophages isolated from *CNN2* KO mice migrated faster than WT controls (Huang et al., 2008).

Calponin 2 regulates cell migration differently in different cell types and in different biological activities. Forced expression of calponin 2 in vascular endothelial cells enhanced angiogenic cell migration in vivo and *CNN2* anti-sense RNA reduced chemotaxis of human umbilical vein endothelial cells in culture (J. Tang et al., 2006). These effects are opposite to the inhibitory effects on cell motility seen in many other cell types. A hypothesis to be tested is that a proper level of calponin 2 is required in balancing the actin cytoskeleton stability versus dynamics in different cell types and in different biological processes, corresponding to a specifically differentiated internal set point of the equilibrium of calponin 2 levels in each case (Hossain et al., 2016; Rasmussen & Jin, 2024).

Significant amounts of calponin 2 are found in human and mouse platelets (Hines et al., 2014). Platelet adhesion is a critical initial step in blood coagulation and thrombosis. A microfluidic thrombosis study found that the time to the initiation of rapid platelet accumulation and thrombosis was significantly longer for blood from *CNN2* KO mice than WT control (Hines et al., 2014). The effect of calponin 2 on facilitating the velocity of cell adhesion was also shown when prostate cancer cells expressing high or low calponin 2 were compared (Moazzem Hossain et al., 2014).

Showing calponin 2 modulation of macrophage motility and inflammatory activation, bone marrow-differentiated monocytes from *CNN2* KO mice proliferated faster than WT control. Calponin 2-null macrophages migrate faster and exhibit enhanced phagocytosis (Huang et al., 2008). In myeloid cell-specific *CNN2* KO mice, the development of inflammatory arthritis was effectively attenuated (Huang et al., 2016). Deletion of calponin 2 in macrophages also significantly attenuated the development of atherosclerosis in apolipoprotein E KO mice (R. Liu & Jin, 2016b).

Demonstrating the underlying mechanism by which calponin 2 regulates the stability, dynamics and functions of the actin cytoskeleton (Hossain et al., 2005), fibroblasts isolated from *CNN2* KO mice show increased cell traction force generated by myosin II motors (Hossain et al., 2016). Consistent with the established role of calponin as an inhibitory regulator of smooth muscle actomyosin ATPase and contractility (Walsh, 1991), the primary mechanism by which calponin 2 regulates actin cytoskeleton functions and dynamics is through tuning myosin motor activities.

2.4. Calponin 2 Determines the Rate and Fate of Cytokinesis

Among the roles of calponin 2 in regulating various cellular functions in different cell types, a biological effect across cell types is the inhibition of cell division. Over-expression of calponin 2 inhibited the rate of cell proliferation (Hossain et al., 2003). Significant amounts of calponin 2 are found in growing smooth muscle tissues of mouse embryos and the uterus during early pregnancy. The expression of calponin 2 decreases to lower levels in adult smooth muscle cells while the expression of calponin 1 is upregulated (Hossain et al., 2003). The expression patterns suggest that as an inhibitory regulator of myosin motor-driven cell motility, a stronger calponin 2 inhibitory regulation acts as a balancing factor in rapidly proliferating cells to effectively regulate dynamic changes of the actin cytoskeleton during cell proliferation especially the process of cytokinesis (Qian et al., 2021).

Supporting this notion, prostate cancer cells with decreased expression of calponin 2 and fibroblasts isolated from *CNN2* KO mice showed significantly increased rates of proliferation than high calponin 2 control cells whereas boosting the level of calponin 2 in prostate cancer cells by transfective expression reduced the rate of cell proliferation (Moazzem Hossain et al.,

2014). Consistently, transfective over-expression of calponin 2 in a rabbit smooth muscle cell line that lack endogenous calponin hindered cell proliferation with an increased number of binuclear cells indicating a blockade of cytokinesis (Hossain et al., 2003).

It is important noting that the functions of calponin 2 in regulating cytokinesis include a rapid proteolytic degradation before the telophase of the cell cycle. While calponin 2 is detected at a high level in cells blocked at the metaphase with colchicine or nocodazole, a rapid and nearly complete proteolytic removal of calponin 2 triggered by PKC phosphorylation is necessary for progressing and completion of cytokinesis. Unregulated viral promoter-driven transfective over-expression of calponin 2 at levels exceeding the proteolytic capacity of the cell blocked cytokinesis and resulted in apoptosis (Qian et al., 2021).

2.5. Role of Calponin 3 in Embryonic Development, Neuronal Plasticity and Cell Fusion

Calponin 3 is present in the brain with a potential role in the functions of actin cytoskeleton during neuronal remodeling (Rami et al., 2006). Calponin 3 was also found in dendritic spines of adult hippocampal neurons to regulate dendritic spine plasticity (Ferhat et al., 2003). While mice with systemic KO of the *CNN1* gene (Takahashi et al., 2000) or *CNN2* gene (Huang et al., 2008) or both (Feng et al., 2019) survive to adulthood and fertile, systemic KO of *CNN3* in mice resulted in embryonic and neonatal lethality due to defects in the development of central nervous system (Flemming et al., 2015). Similarly, *XclpH3*, the *Xenopus* homolog of *CNN3*, was involved in preventing cell's convergence extension movements regulated by *Otx2* (orthodenticle homeobox 2) that is a homeobox gene that plays a role in embryonic brain and systemic development (R. Morgan et al., 1999).

CNN3 is found in the trophoblasts of human placenta and plays a role as a negative regulator of trophoblast fusion. Knockdown of *CNN3* gene expression or dissociation of calponin 3 protein from cytoskeleton in response to PKC phosphorylation promoted trophoblast fusion (Shibukawa et al., 2010). Calponin 3 is also present in myoblasts as an inhibitory regulator of cell fusion. Overexpression of calponin 3 in C_2C_{12} myoblasts inhibited the formation of myotubes during in vitro differentiation whereas knockdown of *CNN3* gene expression promoted fusion and myogenic differentiation. The inhibitory effect of calponin 3 on the cell motility-related functions was reversable with phosphorylation by Rho-associated kinase 1/2 (ROCK1/2) (Shibukawa et al., 2013). These experimental data indicate a critical role of calponin 3 in embryonic development, neural plasticity and myogenesis.

Calponin 3 is found in stress fibers in skin fibroblasts and myofibroblasts in the proliferation phase of wound healing. Supporting the hypothesis that proper level of calponin is required in balancing cytoskeleton dynamics in different cell types and in different biological processes, *CNN3* knockdown in primary fibroblasts in culture impaired stress fiber formation, resulting in decreased cell motility and ability of contraction (Daimon et al., 2013).

2.6. Cellular Regulation of Calponin Functions

The actin thin filaments are an essential part of the contractile machinery of smooth muscle cells. The non-muscle actin cytoskeleton plays multiple vital functions in various cellular activities. Therefore, regulation of the roles of calponin in the actin filament-based cellular functions is important in many physiological processes.

There is a rich collection of in vitro experimental evidences for the phosphorylation regulation of calponin (R. Liu & Jin, 2016a). Data have shown that the residues Ser_{175} and Thr_{184} of calponin 1 are phosphorylated by PKC (Naka et al., 1990). Direct associations of calponin 1 and PKCα (Somara & Bitar, 2008) and PKCε (B. Leinweber et al., 2000) were found in living smooth muscle cells. The primary phosphorylation sites Ser_{175} and Thr_{184} are in the region containing the second actin-binding site (Figure 4). These two PKC substrate residues and the flanking sequences are conserved in all three calponin isoforms, suggesting that similar phosphorylation signaling pathways may also regulate the functions of calponin 2 and calponin 3. It is worth noting that these phosphorylation sites are in the first of the three repeating motifs in the middle region of calponin (Figure 4). The downstream repeats 2 and 3 contain sequences similar to that of the second actin-binding site, including the potential PKC substate serine and threonine residues, which merit further research on their functional significances.

Calmodulin-dependent kinase II and Rho-kinase are also found to phosphorylate calponin in vitro at Ser_{175} and Thr_{184} with Ser_{175} being the main site (Kaneko et al., 2000; Walsh, 1991). Dephosphorylation of calponin as a counter regulation is catalyzed by type 2B and type 1Delta protein phosphatase (Fraser & Walsh, 1995; Ichikawa et al., 1993). An interesting notion was that calponin may activate an autophosphorylation of PKC in a lipid-independent manner (H. R. Kim et al., 2013; B. Leinweber et al., 2000).

The biochemical function of calponin as an inhibitory regulator of myosin ATPase forms a foundation for the broad physiological functions of calponin isoforms, of which the phosphorylation-dephosphorylation regulation may also be a common mechanism. The effects of calponin regulated cytoskeleton force generated by myosin motors on determining the stability and dynamics of cellular structure and motility underlie the role of calponin in various physiological and pathological processes.

Chapter 3

Structure-Function Relationships of Calponin

Primary structures of the three vertebrate calponin isoforms have been determined in multiple species via cDNA cloning and sequencing. The current knowledge of the structure-function relationships of calponin was largely learned from biochemical studies of calponin 1 in the context of smooth muscle functions. Outlined in Figure 4 in Chapter 2, the primary structures of the three calponin isoforms consist of a conserved N-terminal CH domain, a conserved middle region with two actin-binding sites, and a C-terminal variable region that constitutes the size and charge differences of the three isoforms.

3.1. The N-Terminal CH Domain

A sequence motif of ~100 amino acids in the N-terminal region of calponin (e.g., residues 29-129 of calponin 1) is identified as the "calponin homology" (CH) domain. This nomenclature was based on findings of potentially homologous sequence motifs in a number of actin-binding proteins, including α-actinin, β-spectrin, filamin, dystrophin (Gimona et al., 2002) and plectin (Fontao et al., 2001). CH motifs are present in tandem repeats in some proteins.

The CH domain in many proteins has been identified with functions of either a part of actin-binding site or serving as a regulatory structure for the actin binding activity. However, no definitive function has been demonstrated for the CH domain in calponin. The CH domain is not the binding site between calponin and F-actin and does not regulate the modes of calponin-actin binding (Galkin et al., 2006). The CH domain in calponin was found to bind to extra-cellular regulated kinase (ERK) (B. D. Leinweber et al., 1999) and calponin was co-immunoprecipitated with mitogen-activated protein kinase (MAPK) (Menice et al., 1997), implicating a role for calponin to serve as an adaptor protein in the ERK signaling cascade in smooth muscle and non-muscle cells.

3.2. Protein-Binding Sites in the Conserved Middle Region

Extensive in vitro protein binding studies have delineated key functional domains of the calponin family proteins. Two microtubule-binding sites have been identified in calponin, one in the inhibitory segment (residues 145-182) and one in the segment of residues 183-292 (Fattoum et al., 2003). A Ca^{2+}-calmodulin-binding site is in the segment of residues 52-144 (Mezgueldi et al., 1992). A study also reported a heat shock protein 90-binding site, which may function to regulate actin bundle formation, in the N-terminal portion of calponin (residues 7-144) (Y. Ma et al., 2000).

The primary function of calponin is an actin-binding protein that inhibits the Mg-ATPase activity of myosin (Abe et al., 1990; Mezgueldi et al., 1992; Winder, Walsh, et al., 1993; Winder & Walsh, 1990). This inhibition slows down the movement of actin filaments over immobilized myosin heads (Haeberle, 1994; Shirinsky et al., 1992). The actomyosin ATPase inhibitory domain (residues 145-182) in the middle region of calponin binds F-actin through two binding sites (residues 144-162 and 171-188 in chicken calponin 1) and also binds calmodulin (Fattoum et al., 2003). These two actin-binding sites bind F-actin in similar manner and are conserved in all three calponin isoforms (see Figure 4 in Chapter 2).

The effect of calponin on inhibiting actomyosin ATPase is via actin-binding at site 144-162. The site 171-188 flanking Ser_{175} is important for binding to actin, tropomyosin, and calmodulin. Residue Ser_{175} in the actin-binding site 2 is crucial for the inhibitory regulation of myosin ATPase. PKC phosphorylation sites, likely common for all three calponin isoforms, are located at Ser_{175} and Thr_{184} in the second actin-binding site in the first CLIK motif of calponin 1. Calponin 2 may have an additional phosphorylatable serine at residue 177 near the PKC-phosphorylated Ser_{175}. Phosphorylation of calponin at Ser_{175} reverses the inhibition of myosin ATPase (Mino et al., 1998). Phosphorylation of Ser_{175} by PKC reduces the binding affinity of calponin for F-actin, likely through changing the molecular conformation of calponin (J. P. Jin et al., 2000) as a mechanism to diminish the inhibition of myosin ATPase (D. C. Tang et al., 1996).

Calponin increases actin polymerization and inhibits depolymerization of actin filaments. Presence of calponin significantly slows down actin depolymerization (Kake et al., 1995). The two actin binding sites of calponin share their binding sites on actin, involving residues 18-28 and 360-372 of actin. Binding to the 18-28 site of actin results in a reduction in F-actin depolymerization, correlative to the fact that this site of actin determines the release of ATP from actin (Ferjani et al., 2010). This observation is consistent with the effect of calponin 2 on stabilizing actin cytoskeleton in living cells (Hossain et al., 2005).

In addition to binding F-actin, the region of residues 144-188 of calponin is also essential for the binding to tropomyosin and Ca^{2+}-calmodulin (Mezgueldi et al., 1995). High-resolution structural studies of this region indicated that the sequentially located segments 140-146, 159-165, 189-195, and 199-205 tend to form four transient α-helices, likely providing the conformational malleability needed for the functionally promiscuous nature of this region of calponin (Pfuhl et al., 2011).

There are three repeating sequence motifs in calponin (the CLIK motifs, see Figure 4 in Chapter 2). This repeating structure is conserved in all three isoforms and across species. The first repeating motif overlaps with the second actin-binding site (residues 171-188) and contains the PKC substrate residues Ser_{175} and Thr_{184} that lack counterparts in the first actin-binding site (residues 144-162). This structural feature is consistent with the hypothesis that the second actin-binding site plays a regulatory role in the binding of calponin to the actin filament (Mino et al., 1998). Similar sequences as well as potential phosphorylation sites are present in repeats 2 and 3 but their function and regulation have not been determined. Therefore, the biological significance of these repeating motifs in calponin remains to be established.

3.3. The C-Terminal Variable Region

The C-terminal segment of calponin is a variable region that has diverged significantly among the three isoforms. The variable lengths and amino acid compositions of the C-terminal segment produce the size and charge differences among the three calponin isoforms, defining the overall basic, neutral, and acidic charge property of the isoforms, which have rendered calponin 1, 2 and 3 the previous used nomenclatures of "basic," "neutral" and "acidic" calponins, respectively (J.-P. Jin et al., 2008; R. Liu & Jin, 2016a; Wu & Jin, 2008).

The C-terminal variable segments of calponin isoforms have been shown to have different effects on weakening the binding of calponin to F-actin. Deletion of the C-terminal tail segment of calponin strongly enhanced the actin-binding and bundling activities of all three isoforms, indicating a regulatory function of the C-terminal variable region (Bartegi et al., 1999; Danninger & Gimona, 2000). Domain-swapping experiments demonstrated that having the C-terminal segment of calponin 2 decreased the association of all three isoforms to actin cytoskeleton whereas the C-terminal segment of calponin 1 had little effect. Consistently, calponin 2 appears having the lowest binding affinity for F-actin among the three isoforms. The C-terminal tail of calponin regulates the interaction with F-actin by altering the function of the second actin-binding site in the middle region (Burgstaller et al., 2002). These findings suggest that C-terminal variable region may determine cell type-specific functions and/or subcellular distributions of the calponin isoforms.

3.4. High Resolution 3-D Structure and Actin Filament Associated Structure of Calponin

To date, no high-resolution folded structure of intact calponin has been determined. Nuclear magnetic resonance (NMR) solution structure of the CH domain has been determined in calponin 1 (Molecular Modeling Database IDs 34608 and 18703, Protein Data Bank accession #s 1WYP and 1H67) and calponin 2 (Protein Data Bank accession #s 1WYN and 1WYM). Fitting the NMR structure of calponin CH domain to the 3D-helical reconstruction of F-actin-calponin complex of cryo-electron microscopic images (Bramham et al., 2002) provided a structural insight into the mode of interaction between F-actin and calponin and other CH domain-containing proteins.

In another electron microscopic 3-D reconstruction study of the structural organization of calponin on actin and actin-tropomyosin filaments (Hodgkinson et al., 1997), calponin density was found to occur peripherally along the long-pitch of the F-actin helix. The main calponin mass was located over sub-domain 2 of actin and connected axially adjacent actin monomers by binding to the "upper" and "lower" edges of sub-domains 1 of each actin monomer. When the reconstructions were fitted to the atomic model of F-actin, calponin appeared to contact actin near the N terminus and at residues 349 to 352 close to the C terminus in sub-domain 1 of one monomer and residues 92 to 95 in sub-domain 1 and residues 43 to 48 in sub-domain 2 of the axially neighboring monomer. These positions are also binding sites for some other actin-associated proteins and are near the sites for weak myosin interactions. When added to tropomyosin-containing actin filaments, calponin caused a shift of tropomyosin away from sub-domain 1 towards sub-domain 3 of actin, exposing strong myosin-binding sites that were previously covered by tropomyosin. This structural effect is unlike that of troponin in striated muscle thin filament and therefore the regulatory inhibitions of actomyosin ATPase by calponin and troponin may not be analogous.

Transgelins are members of the calponin family proteins (T.-B. Hsieh & Jin, 2023). Transgelin contains the core structure shared by the three calponin isoforms but has only one actin-binding site without the repeating motifs and the C-terminal tail (R. Liu et al., 2017). High-resolution crystal structure of full length human transgelin-1 (SM22α) has been obtained (M. Li et al., 2008), showing a conserved CH domain similar to the CH domains in other proteins. The high-resolution structure of transgelin with less complex actin-binding and regulatory structures may provide valuable information for understanding the biochemically determined functional regions of calponin.

To further establish the structure-function relationship of calponin and the three isoforms will be an essential step toward fully understanding calponin's physiological functions in different cell types and various physiological and pathological conditions. Even though calponins are abundant proteins in smooth muscle and many other cell types with critical roles in various cellular functions and changes of expression level in various tumor cells as detailed in flowing chapters, no human disease-causing mutations have been reported to date. This information gap suggests a hypothesis that calponin may have a high tolerance to point mutations unless the mutations are at critical sites to cause embryonic lethality. Further research on the structure-function relationships of calponin will ultimately answer this intriguing question.

Chapter 4

Double Deletions of Calponin 1 and Calponin 2 Decrease Blood Pressure and Blunts Myogenic Response of Vascular Smooth Muscle

Calponins are abundant in smooth muscle cells. Calponin 1 is specifically expressed in differentiated and contractile smooth muscle and has been extensively studied for its role in the regulation of smooth muscle contractility (K. G. Morgan & Gangopadhyay, 2001; Small & Gimona, 1998; Winder et al., 1998). Based on the biochemical function of calponin as an actin filament associated regulator of smooth muscle myosin ATPase (Winder et al., 1998), calponin 1 has been shown to modulate smooth muscle contractility (Matthew et al., 2000; Nigam et al., 1998). Smooth muscles also contain an abundant amount of calponin 2 (Hossain et al., 2003). Calponin 1 and calponin 2 have largely conserved primary structures (J.-P. Jin et al., 2008; R. Liu & Jin, 2016a). Besides its role in regulating cytoskeleton functions and cell motility as shown in various non-muscle cell types and organ systems (R. Liu & Jin, 2016a; Rasmussen & Jin, 2024), calponin 2 shows regulated expression during smooth muscle development and remodeling (Hossain et al., 2003) and also functions in modulating smooth muscle contractility (Feng et al., 2019). Therefore, it is more informative to investigate the functions of calponin 1 and calponin 2 in regulating smooth muscle contractility together in an integrative experimental system using in vivo and ex vivo approaches.

4.1. *CNN1*$^{-/-}$ and *CNN2*$^{-/-}$ Double Knockouts Do Not Produce Lethality in Mice

Mice of *CNN1* KO (Matthew et al., 2000) or *CNN2* KO (Huang et al., 2008) survive, fertile and have normal life spans. These single KO mice exhibit only minimum smooth muscle phenotypes. Therefore, calponin 1 and calponin 2 may be functionally complementary in smooth muscle cells at least partially. To investigate the potentially overlapping functions of calponin 1 and calponin

2 in smooth muscle, the *CNN1*-null and *CNN2*-null alleles were combined by cross breeding to create double KO mice (Feng et al., 2019). The result showed that systemic *CNN1*,*CNN2* double knockouts in mice do not cause lethality and the *CNN1*$^{-/-}$,*CNN2*$^{-/-}$ mice have normal life span in standard cage conditions to provide an experimental system to examine the effects on smooth muscle functions.

4.2. *CNN1*$^{-/-}$,*CNN2*$^{-/-}$ Mice Show Decreased Systemic Blood Pressure

To evaluate the effect of deletions of both calponin 1 and calponin 2 on vascular smooth muscle function in vivo, blood pressure measurement using tail-cuff in conscious *CNN1*$^{-/-}$,*CNN2*$^{-/-}$ mice of C57BL/6 background detected significantly lower systolic (88.3 ± 2.6 mmHg vs 106.9 ± 3.7 mmHg, mean ± SEM), diastolic (61.6 ± 3.1 mmHg vs 77.8 ± 2.8 mmHg), and mean (70.5 ± 2.9 mmHg vs 87.5 ± 3.0 mmHg) arterial blood pressures in comparison with WT controls ($p < 0.05$) (Feng et al., 2019). In contrast, *CNN1* or *CNN2* single KO mice exhibited only trends of decrease in blood pressures, more visible for the diastolic pressure. *CNN1* single KO showed more notable effect than that of *CNN2* single KO, consistent with the fact that calponin1 is smooth muscle-specific and the major calponin isoform in smooth muscle cells. On the other hand, the additive effects of *CNN1*$^{-/-}$,*CNN2*$^{-/-}$ double KO on lowering blood pressure indicate that calponin 1 and calponin 2 both contribute to the physiological tone of arterial smooth muscle in regulating blood pressures in vivo. This phenotype of *CNN1*$^{-/-}$,*CNN2*$^{-/-}$ double KO mice demonstrate the function of calponin in modulating vascular smooth muscle contractility with physiological and pathological significances.

4.3. *CNN1*$^{-/-}$,*CNN2*$^{-/-}$ Double KO Alters Vascular Smooth Muscle Contractility

Increased vascular smooth muscle relaxation (Rinaldi, 2005) and vasodilation (Ferrari et al., 2001; Fritz & Rinaldi, 2007) reduce blood pressure whereas impaired relaxation of arterial smooth muscle causes hypertension (Qiao et al., 2014). *CNN1*$^{-/-}$ mice were found with impaired blood pressure responses during exercise due to increased vasodilation (Masuki, Takeoka, Taniguchi,

& Nose, 2003; Masuki, Takeoka, Taniguchi, Yokoyama, et al., 2003). Showing a role of calponin 1 in modulating smooth muscle contratility, vas deferens smooth muscle of *CNN1*$^{-/-}$ single KO mice has faster contractile velocity as compared with that of WT control (Matthew et al., 2000; Takahashi et al., 2000). On the other hand, high extracellular KCl-induced isometric force was lower in *CNN1*$^{-/-}$ single KO mouse vas deferens and aortic smooth muscles (Fujishige et al., 2002). *CNN1*$^{-/-}$ single KO mice exhibited blunted blood pressure response to phenylephrine stimulation than WT control (Masuki, Takeoka, Taniguchi, & Nose, 2003; Masuki, Takeoka, Taniguchi, Yokoyama, et al., 2003), similar to that of isolated aortic smooth muscle ring of Wistar Kyoto rats, which natually lacks calponin (Nigam et al., 1998).

Demonstrating the effect of deleting both calponin 1 and calponin 2 on arterial smooth muscle contractility, endothelial-free aortic rings from *CNN1*$^{-/-}$,*CNN2*$^{-/-}$ double KO mice showed lower norepinephrine activated maximum tension development than that of WT control (3.43 ± 0.28 mN/mm^2 vs 6.04 ± 0.32 mN/mm^2 at 1 μM norepinephrine, mean ± SEM, $p < 0.001$) (Feng et al., 2019). Consistently, it has been observed that *CNN1*,*CNN2* double KO female mice have dystocia, indicating weakened uterus contractility (our unpublished observation). The lower maximum force seen in *CNN1*$^{-/-}$,*CNN2*$^{-/-}$ double KO mouse aortic rings was also produced by *CNN1*$^{-/-}$ single but not *CNN2*$^{-/-}$ single knockout mouse aortae, indicating that the deletion of calponin 1, but not calponin 2, predominantly contributed to the reduction of force development in aortic smooth muscles. *CNN1*$^{-/-}$ single KO but not *CNN2*$^{-/-}$ single KO mouse aortic rings further exhibited shorter relaxation time similar to that of *CNN1*$^{-/-}$,*CNN2*$^{-/-}$ double KO group (Feng et al., 2019). On the other hand, the norepinephrine dose response curve of *CNN1*$^{-/-}$,*CNN2*$^{-/-}$ double KO mouse aortic rings was unchanged, indicating that the effect of *CNN1*$^{-/-}$,*CNN2*$^{-/-}$ double KO on vascular smooth muscle contractility is downstream of the norepinephrine-induced Ca^{2+} activation.

4.4. *CNN1*$^{-/-}$,*CNN2*$^{-/-}$ Double KO Blunted Length-Active Tension Response of Aortic Smooth Muscle

Blood pressures are determined by vessel tone dynamically maintained by tonic contractions of vascular smooth muscle. While the lower maximum force and faster relaxation of *CNN1*$^{-/-}$,*CNN2*$^{-/-}$ double KO mouse smooth muscle reflect altered vascular tone in the regulation of blood pressure, they do not directly relate to lower blood pressures. The myogenic response (Folkow, 1989) serves as a fundamental mechanism in the function of arterial smooth muscle, in which a positive resting length-tension relationship is a determining factor (Walsh & Cole, 2013). The passive property of smooth muscle is an important factor in determining contractility especially in the myogenic response of resistant blood vessels (Folkow, 1989). While the resting tension was unchanged, the length-active tension response of *CNN1*$^{-/-}$,*CNN2*$^{-/-}$ double KO mouse aortic smooth muscle was blunted (Feng et al., 2019), suggesting that calponin plays a critical role in myogenic response of vascular smooth muscle.

Mechanical tension transduced to the vascular smooth muscle cytoskeleton is a basis of the myogenic response (Davis et al., 2001). The blunted length-active tension response in *CNN1*$^{-/-}$,*CNN2*$^{-/-}$ double KO mouse aortae is not seen in the *CNN1*$^{-/-}$ or *CNN2*$^{-/-}$ single KO model. These phenotypes suggest that the combination of calponin 1's function in directly tuning smooth muscle contractility and calponin 2's function in regulating cytoskeleton myosin motors and tension responses (Hossain et al., 2005, 2016) is important for the myogenic response of arterial smooth muscle.

A study of *CNN1*$^{-/-}$ mice showed that deletion of calponin 1 alters aortic smooth muscle contraction and relaxation with no alteration of the length-resting tension relationship (Fujishige et al., 2002). With its abundance in smooth muscle cells, calponin 2 may play a role in the vascular sensing of passive tension in myogenic response. While the deletion of calponin 1 lowers the length-dependent force generation, the deletion of calponin 2 increases cytoskeleton myosin motor activity (Hossain et al., 2016) to increase the dynamics of the cytoskeleton and damp the stretching signals which results in decreased sensing and myogenic response of the contractile machinery. Supporting the role of calponin 2 in myogenic response as a potential target for the treatment of hypertension, aortic smooth muscle expresses a much higher level of calponin 2 relative to that of calponin 1 in contrast to the high calponin 1 to calponin 2 ratios in urinary bladder and large intestinal smooth muscles (Feng et al., 2019).

4.5. Compensatory Increase of Calponin 1 in *CNN2*$^{-/-}$ Mouse Aortic Smooth Muscle

Normalized to the level of actin and myosin, the amount of calponin 1 was significantly increased in *CNN2*$^{-/-}$ mouse aortae in comparison to the level in WT control where significant amount of calponin 2 is present. This finding indicates a compensation for the loss of calponin 2 with increased synthesis or accumulation of calponin 1. No such compensatory increase of calponin 2 was found in *CNN1*$^{-/-}$ mouse aortic smooth muscle (Feng et al., 2019). This unidirectional compensation between calponin 1 and calponin 2 was also seen in urinary bladder and large intestinal smooth muscles of *CNN1*$^{-/-}$ and *CNN2*$^{-/-}$ single KO mice (Feng et al., 2019). A possible reason for the unidirectional compensatory change is that the expression of calponin 2 in mouse aortic smooth muscle and likely in other smooth muscles may be predominantly regulated by an intrinsic set point determined by cell and tissue mechanical environments (Rasmussen & Jin, 2024), further supporting the potential value of targeting calponin 2 for the treatment of hypertension (Feng et al., 2019).

4.6. Functional Diversity and Exchangeability between Calponin Isoforms

The primary structure of the three calponin isoforms are highly conserved in the N-terminal and middle regions that contain all known core functional sites (see Figure 4 in Chapter 2 and R. Liu & Jin, 2016a). While the isoform-specific C-terminal variable region in calponin 1 and calponin 2 may determine their subcellular distribution and functional diversities, the compensatory increase of calponin 1 in calponin 2-null mouse smooth muscle suggests a functional exchangeability of these two isoforms. On the other hand, the very different ratios of calponin 1 and calponin 2 in aorta, bladder and large intestine may correspond to different smooth muscle types with diverged functional features, such as tonic and phasic contractions. The calponin 1 compensation for the loss of calponin 2 in smooth muscle of *CNN2*$^{-/-}$ single KO mice may also have effects on modifying contractile phenotypes as a secondary adaptation, which is worth further investigating.

The other isoform of calponin, calponin 3, is also expressed in smooth muscle but at low levels (Applegate et al., 1994). The physiological function of calponin 3 in smooth muscle contractility is less understood than that for calponins 1 and 2. The *CNN1*$^{-/-}$,*CNN2*$^{-/-}$ double knockout mice may provide an informative experimental system to examine the function of calponin 3 in smooth muscle contractility in the absence of the other two isoforms.

The decreased blood pressure of *CNN1*$^{-/-}$,*CNN2*$^{-/-}$ double KO mice demonstrates that calponins 1 and 2 have a combined novel role in blood pressure regulation via adjusting myogenic response of resistant blood vessels. Epidemiology data published by the US Center for Disease Control and Prevention on hypertension in 2020 (NCHS Data Brief No. 364) showed that in 2017–2018, the prevalence of age-adjusted hypertension was 45.4% among all adults. The finding that double deletion of calponins 1 and 2 in arterial smooth muscle effectively lowers blood pressure suggests that co-suppression of the expressions or functions of calponin 1 and calponin 2 in vascular smooth muscle can be explored as a myofilament mechanism to develop a new treatment for hypertension, which has the advantage of targeting the contractile regulation downstream of neuronal and hormonal signaling pathways with potentially less systemic side effects.

Chapter 5

Deletion of Calponin 2 Results in Low Numbers and Premature Depletion of Ovarian Follicles

Calponins are abundant proteins expressed in multiple cell types. The functional importance of calponin is shown by the embryonic and neonatal lethality of *CNN3* KO mice. However, *CNN1*, *CNN2,* or *CNN1*,*CNN2*-double KO mice survive, fertile and have normal life span at least in laboratory cage conditions (Feng et al., 2019). Therefore, it is important to establish the fitness values of the two isoforms of calponin for their evolutionary selection and fixation. This line of evidence includes calponin 1's determining role in smooth muscle contractility, which was reported as a clinical maternal serum marker for short-term prediction of preterm birth (Cetin et al., 2018). Phenotype studies of *CNN2* KO mice have provided plausible leading information for the biological importance of calponin 2. An example is that calponin 2 expression is detected in the ovary (Hossain et al., 2006) and it regulates cell proliferation (Hossain et al., 2003; Qian et al., 2021), suggesting a role in ovarian development and function to determine the reproductive capacity during natural selection.

Ovarian reserve established during fetal development is the major factor determining the longevity of ovarian function. Besides intraovarian paracrine and humoral factors, mechanical tension plays a role in controlling the selection and maintaining dormant state of oocyte in primordial follicle in mice (Nagamatsu et al., 2019; Shah et al., 2018). In vertebrates, the ovary is enclosed in a layer of tunica albuginea made of dense collagen fiber, which maintains a constantly fluctuating intraovarian pressure. At the initiation of an ovulation cycle, a wave of the ovarian follicles is activated and enlarged in size with rapid proliferation of granulosa cells in response to the stimulation of gonadotropins as well as accumulation of the follicular fluid until ovulation occurs. In human, the size increase from primary follicle of less than 1 mm diameter to a 20 -27 mm final diameter of the Graafian follicles (Coelho Neto et al., 2018) generates a constantly increased intraovarian pressure before ovulation. During this period, a more than 10 folds increase in the volume of growing follicles exerts high pressure to the adjacent small follicles and inhibit the development of other follicles. In this process, calponin 2 regulation of actin cytoskeleton functions may play a role in mediating the mechanoregulation of ovarian follicles.

5.1. Female *CNN2* KO Mice Are Fertile But Produce Litters of Significantly Smaller Size than WT Controls

CNN2 KO mice of both sexes show normal overall growth and cage activities as compared with WT littermates. Homozygous *CNN2* KO mice have normal life expectancy in non-stressed cage environment and could live up to 28 months old (T. B. Hsieh et al., 2022; T. B. Hsieh & Jin, 2022; Huang et al., 2008, 2016; Plazyo et al., 2018). Statistically analyzed quantitative data show no significant differences in body weight between WT and homozygous *CNN2* KO mice at 4-12 weeks old and at 15-17 months old (T.-B. Hsieh & Jin, 2024).

Homozygous and heterozygous *CNN2* KO mice both produce offspring naturally. The offspring genotype ratios show normal patterns of Mendelian segregation with heterozygous parents produced 26.5% homozygous *CNN2* KO, 50.0% heterozygous and 23.5% WT pups. The phenotype indicates that male and female *CNN2* KO gametes retain the ability of fertilization and *CNN2* KO homozygous embryos have normal rates of implantation and development as compared to WT or heterozygous embryos. No sex bias was seen among the offspring. The male to female ratios of pups produced from homozygous, heterozygous and WT parents were similar (1.08, 1.10 and 1.08, respectively, T.-B. Hsieh & Jin, 2024).

A striking finding was that homozygous female *CNN2* KO mice produced significantly smaller litter sizes than WT control (both are of C57BL/6 background), indicating a lower reproductive performance. The average litter size produced by homozygous *CNN2* KO female mice at ages of 1-9 months old was 5.08 ± 1.98 (mean ± SD), significantly smaller than that produced by age-matched WT female mice (8.31 ± 1.44), $p < 0.001$ (T.-B. Hsieh & Jin, 2024). The litter size from heterozygous dams was 8.04 ± 1.64, not different from the WT control. Homozygous *CNN2* KO male paired with WT female mice produced an average liter size of 9.21 ± 1.53, no less than control breeding produced by age-matched WT males, demonstrating that *CNN2* KO does not affect male reproductivity. The breeding data concluded that the loss of calponin 2 specifically decreases female reproductivity.

5.2. Homozygous Female *CNN2* KO Mice Demonstrate an Age-Progressive Decline of Litter Size

Analysis of the trends of litter sizes produced by 1-9 months old female *CNN2* KO and WT mice demonstrated that homozygous *CNN2* KO females have an age-progressive significant declining in litter sizes while the WT controls remained stable during this time period. It is worth noting that homozygous *CNN2* KO female mice produced offspring at the same initial age of reproduction (one month old) as that of WT female mice, demonstrating that the deletion of calponin 2 does not affect the onset of puberty. The similar litter sizes produced by one month old female *CNN2* KO and WT mice further indicate that *CNN2* KO does not impair the systemic capacity of carrying normal number of embryos at young age (T.-B. Hsieh & Jin, 2024). With the age-progressive declining in litter size, *CNN2* KO females exhibit early depletion of ovarian follicles. While 15-17 months old WT mouse ovaries still contained hundreds of ovarian follicles, the ovaries of *CNN2* KO mice at the same age had no follicles detected (T.-B. Hsieh & Jin, 2024). These results show that the loss of calponin 2 decreases the life expectancy of ovarian reserve.

5.3. Young *CNN2* KO Female Mice Show Significantly Lower Total Ovarian Follicle Counts and High Numbers of Atretic Follicles than WT Controls

Total ovarian follicle count is the most important factor determining the life expectancy of ovarian reserve. Histological examination of serial sections of whole ovaries of homozygous *CNN2* KO and WT female mice in young and older age groups to compare their ovarian follicle counts found that homozygous *CNN2* KO mice had significantly lower total follicle counts than WT controls at 2 months (2,988 ± 221 vs 5,280 ± 694, $p < 0.05$), 9 months (546 ± 99 vs 1,062 ± 190, $p < 0.05$) and 16 months (87 ± 21 vs 261 ± 55, $p < 0.05$) old. It is important to note that the numbers in 2 months old young *CNN2* KO female mice are less than 60% of the WT control, indicating impaired ovarian folliculogenesis and low ovarian reserve. The difference between *CNN2* KO and WT groups became larger at 9 and 16 months of age, demonstrating that *CNN2* KO causes a faster progressive loss of ovarian follicles (T.-B. Hsieh & Jin, 2024).

In homozygous young (2 months old) female *CNN2* KO mice, the number of both single-layer and multi-layer ovarian follicles were significantly lower than that of age-matched WT littermates (T.-B. Hsieh & Jin, 2024). The number of single layered follicles indicates the major ovarian follicle pool established in the prenatal period and represents the ovarian reserve and the longevity of ovulatory cycles. Multi-layered ovarian follicles represent the follicles recruited and activated for further development. The finding that female *CNN2* KO mice had significantly lower numbers of single layered follicles demonstrates a role of calponin 2 in folliculogenesis. The poor ovarian reserve of *CNN2* KO mice may owe inadequate primordial follicle formation and/or excessive ovarian follicle apoptosis in a prenatal stage.

Compared to WT control, the ovaries of young *CNN2* KO mice also had higher numbers of atretic follicles. Most of the atretic follicles displayed collapsed oocytes with only empty zona pellucida remaining. Abnormal follicles such as multi-oocyte follicles (MOF) were also present in the ovaries of young *CNN2* KO mice (T.-B. Hsieh & Jin, 2024). MOFs have been reported in several species and proposed to result from abnormal nest breakdown, dysregulated follicle assembly and/or incomplete ovigerous cord breakdown. The existence of MOFs can be used as a marker of abnormal folliculogenesis (Gaytán et al., 2014). Several transgenic animal models have shown increased MOFs, such as mice with a knockout of the BMP15/GDF-9 (Yan et al., 2001), Notch-2 (Xu & Gridley, 2013) or FSH/inhibin (T. R. Kumar et al., 1999) gene. Animal studies also found that fetal exposure to environmental toxins, such as bisphenol A and genistein, or irradiation (Mazaud Guittot et al., 2006; Rodríguez et al., 2010) leads to granulosa cell migration or assembly with oocyte to promote the formation of MOFs. The effect of calponin 2 deletion on the formation of MOFs could also be via other granulosa cell dysfunctions and is worth further investigating (T.-B. Hsieh & Jin, 2024).

5.4. Calponin 2 Is Expressed in Oocytes and Cumulus Cells

Calponin 2 is known to express in multiple cell types such as skin keratinocytes, fibroblast, epithelial cells, endothelial cells and macrophages (R. Liu & Jin, 2016a). Immunofluorescence stain of cumulus-oocyte-complex collected from the oviducts of a young female WT mouse showed that calponin 2 is expressed in the cytoplasm of oocyte and the surrounding cumulus cells. Female *CNN2* KO mice are fertile with no significant decrease in litter size at 4-8 weeks old. Therefore, the loss of calponin 2 in oocytes is unlikely a primary cause of ovarian follicle atresia. Reflecting the expression of calponin 2 in cells unrelated to ovarian folliculogenesis and development, immunofluorescence stain of 4-cell cleavage stage embryos collected from a female WT mouse also detected calponin 2 at the peripheral of the blastomeres. The atretic follicles are mostly in secondary or tertiary follicles with few primordial follicles. In secondary or tertiary follicles, oocytes are arrested in the first meiotic division and nutrients need to be supplied by the granulosa cells through transzonal projection. Therefore, the loss of calponin 2 function in granulosa cells may be the major cause of follicular atresia (T.-B. Hsieh & Jin, 2024).

Immunofluorescence microscopy found colocalization of calponin 2 and actin stress fibers and myosin II motors in granulosa cells. An inhibitor of myosin ATPase, calponin 2 in ovarian granulosa cells may function in a physiological equilibrium of myosin motor-based cellular force (Hossain et al., 2016) to maintain cell integrity and connection with neighboring cells. Calponin 2 plays a role in in cell migration (R. Liu & Jin, 2016b) and cytokinesis (Qian et al., 2021). Ovarian granulosa cells are rapidly proliferating from a single layer into multiple layers of the antral follicles. Therefore, cell migration and proliferation may be an underlying mechanism for *CNN2* KO to cause premature ovarian insufficiency (POI). Actin stress fibers are seen in filamentous threads stretched out in the cytoplasm of primary cultures of ovarian granulosa cells, in which calponin 2 colocalizes with actin, tropomyosin and myosin II, implicating a role in regulating cytoskeleton structure and tension-related functions (T.-B. Hsieh & Jin, 2024).

5.5. *CNN2*-KO Mice Provide an Animal Model to Study POI

Reproductive ageing in women includes accelerated declining in ovarian follicle number and egg quality. This natural process occurs progressively and leads to total loss of fecundity, cycle regularity, and eventually menopause ensued at around age 50. POI is defined as menopause occurred before 40 years of age. The exact mechanisms causing primary premature ovarian failure remain unknown (Chon et al., 2021).

To date, there is no ideal animal model for studying the disease process of idiopathic POI. Most of the POI mouse models involve mutations that result in sterilization or loss of primordial germ cells at neonatal or early prenatal period (Duncan et al., 1993; Jagarlamudi et al., 2010; Kõks et al., 2016; Vaz et al., 2020), imposing limitations to studying the natural course of idiopathic

POI. Mice treated with anticancer drugs are alternatives of animal models for studying POI. Although these models are suitable for the study of gonadotoxin-induced POI, they are very different from the clinical scenarios of idiopathic POI (E. H. Lee et al., 2018). In contrast, the reproductive phenotypes of homozygous *CNN2* deletion in mice mirror the clinical scenario of idiopathic POI and poor ovarian reserve in humans. With fertility and a life expectancy similar to that of WT mice, *CNN2* KO mice provide a new platform for longitudinal investigations to study the progression of POI and a new model system to study the role of ovarian granulosa cells in the pathogenesis of idiopathic POI.

5.6. Calponin 2 as a Molecular Target Regulating Intra-Ovarian Mechanical Tension Signaling in Folliculogenesis and POI

Most of the calponin 2 expressing cells reside in a tissue environment involving mechanical tension. Ovarian folliculogenesis is a highly dynamic process and needs intercellular communications between different types of the somatic cells as well as with the oocytes through the transzonal projections. Studies have proposed that intraovarian mechanical tension may play a role in maintaining most of the primordial follicles in resting state (Nagamatsu et al., 2019; Sun et al., 2021). Cyclically developed growing ovarian follicles exert pressure tension in the tissue matrix and resting follicles in the ovary, transmitting pressure signals to the granulosa cells surrounding the oocytes, which are the main structures to absorb the mechanical stress. With abundant actin filaments in granulosa cells, it is likely that calponin 2 may regulate the actin cytoskeleton in sensing the pressure environment of ovarian follicles during folliculogenesis, development and POI.

Three dimensional ultrasound studies showed that the volume of human ovary increases by 70-80% of the original size before ovulation (Jokubkiene et al., 2006). The expansion of the ovary is limited by the tunica albuginea, generating pressure to the interior of the ovary. The non-dominant ovarian follicles are regulated by intraovarian tension. Actin microfilaments is the major cellular structure that stabilizes the delicate intracellular structure to maintain their functional integrity. Actin polymerization-modulating drugs regulate ovarian follicle growth via the Hippo/YAP pathway (Cheng et al., 2015). Modulating actin cytoskeleton dynamics induced disruption of Hippo pathway, leading to the production of CCN growth factors and anti-apoptotic BIRC that are factors activating the growth of secondary follicles (Kawashima & Kawamura, 2018). Therefore, the functions of calponin 2 in regulating mechanical tension in the actin cytoskeleton (Hossain et al., 2005, 2006, 2016; Jiang et al., 2014; Rasmussen & Jin, 2024) may plays a critical role in maintaining the structure and function of granulosa cells that are adapted to a physiologically fluctuating pressure environment. The role of calponin 2 in ovarian functions and female reproduction efficiency not only demonstrates its evolutionary selection value but also presents a new direction for in depth mechanistic studies of POI in humans to develop targeted treatment.

Chapter 6

Loss of Calponin 2 in Kidney Podocytes Results in Age-Progressive Proteinuria

Calponin 2 is expressed in kidney glomerular podocytes at a high level (T. B. Hsieh & Jin, 2022). Chronic kidney diseases affect more than 10% of the general population, and most of the pathogenesis originates from glomerulus injuries (Assady et al., 2019). Regardless of the initial cause of glomerular injury, proteinuria is usually the most common and major manifestation of kidney diseases. While proteinuria may occur transiently in healthy subjects after strenuous exercise or during pregnancy (Weissgerber et al., 2014), persistent and progressive proteinuria usually leads to end stage kidney failure.

Renal filtration occurs in the glomerulus that has a barrier structure consisting of three layers: Fenestrated capillary endothelium, the glomerular basement membrane (GBM) and the podocytes. Podocytes are highly differentiated epithelial cells that cover the outer layer of the glomerular tufts with projected primary and secondary foot processes. The primary foot processes (PFP) attach directly to the GBM through adhesion molecules and interdigitate with neighboring PFPs bridged with slit diaphragm to provide a major support to sustain the integrity of glomerular capillary against the hydrostatic filtration pressure. Actin cytoskeleton in podocytes bears the hydrostatic pressure and high tension in the glomerular filtration assembly (Endlich & Endlich, 2006). By adjusting the tension and stability of actin filaments through inhibitory regulation of myosin II-dependent cell traction force (Hossain et al., 2016), calponin 2 regulates actin-cytoskeleton structure and function to maintain a physiological level of mechanical dynamics and cytoskeleton stability (Jensen et al., 2014). Through this regulatory function, calponin 2 plays a critical role in sustaining the structure of podocytes and glomerular filtration (T. B. Hsieh & Jin, 2022).

6.1. Calponin 2 Is Expressed at High Levels in Kidney Podocytes

Frozen sections of WT mouse kidney stained with an anti-calponin 2 monoclonal antibody (mAb) detected calponin 2 in the glomerular tufts. While podocin expressed linearly in the peripheral edge along the capillary tufts which represent foot processes of the podocyte, calponin 2 is present in the central portion of the capillary loop, corresponding to the bodies of podocytes with continuously lining the vessel tufts (T. B. Hsieh & Jin, 2022).

The high level expression of calponin 2 in renal podocytes was further shown in cultures of a mouse podocyte cell line E11 (T. B. Hsieh & Jin, 2022). After 14 days of differentiation, adherent proliferating E11 podocytes cultured on collagen coated glass cover slips expressing characteristic marker proteins such as podocin and nephrin expressed high levels of calponin 2 as detected in Western blot. Normalized to the level of actin, proliferating podocytes showed higher expression of calponin 2 than that of differentiated podocytes while no difference was found in the levels of podocin (T. B. Hsieh & Jin, 2022).

Immunofluorescence imaging demonstrated the association of calponin 2 with actin stress fibers. For calponin's primary activity as a regulator of myosin motors, its cell body-centralized localization in podocytes may indicate functions in maintaining the myosin motor-based overall cellular tension (Hossain et al., 2016) that determines cell adhesion and the connection with neighboring cells. Immunofluorescence microscopic studies further showed that calponin 2 colocalizes with actin microfilaments, tropomyosin and myosin IIA in E11 podocytes differentiated in culture, demonstrating its association with the actin cytoskeleton in cellular projections (T. B. Hsieh & Jin, 2022). Colocalization of calponin 2 with myosin IIA indicates its function in regulating myosin motor activity and mechanical tension dynamics (Hossain et al., 2016) of the cytoskeleton of kidney podocytes.

6.2. *CNN2* KO Mice Exhibit Age-Progressive Proteinuria

While *CNN2* KO mice have a similar life expectancy as that of WT mice, they exhibit age-progressive proteinuria, indicating chronic deterioration of kidney function (T. B. Hsieh & Jin, 2022). The mean values of urine albumin/creatinine (A/C) ratio (mg/g) of young and old *Cnn2* KO and WT mice were measured as a normalized indicator of proteinuria. No significant difference in A/C ratio was found between *CNN2* KO and WT littermates at 3-8 months old. However, 18-24 months old *CNN2* KO mice showed significantly higher A/C ratio (78.16 ± 31.58, mean ± SD) than that of 3-8 months *Cnn2* KO mice (17.90 ± 5.23) and 18-24 months old WT mice (26.19 ± 5.09) ($p < 0.001$), demonstrating age-related severe albuminuria. From proteins leaking from the glomeruli, a significantly higher number of proteinaceous casts was found in dilated renal tubules of aging *CNN2* KO mouse kidney in comparison to the age matched WT control.

Research has shown that decaying renal function in aging C57BL/6 strain of mice is less likely to produce proteinuria in comparison with other strains studied (Hackbarth & Harrison, 1982; L. J. Ma & Fogo, 2003). Therefore, the age-progressing development of significant proteinuria in the *CNN2* KO mice of C57BL/6 background convincingly demonstrates the critical role of calponin 2 in maintaining the structural integrity and filtration function of renal glomeruli. This finding suggests that the role of calponin 2 in regulating the stability of actin cytoskeleton in adaptation to mechanical tension is a pivotal factor in the pathogenesis of proteinuria. The loss of calponin 2 function results in progressive degeneration and protein leakage of the kidney glomeruli presumably by causing filtration tension-related podocyte injuries. This finding in *CNN2* KO mice urges clinical consideration of including *CNN2* genetic testing for the diagnosis of etiologically unknown proteinuria patients.

6.3. *CNN2* KO Mice Show Age-Progressive High Percentage of Low G/B Ratio Glomeruli

The density of glomeruli was not significantly different between *CNN2* KO and WT mouse kidneys at young adult or old ages. However, aging *CNN2* KO mice (18-24 months old) showed a significantly higher percentage of glomeruli with low glomerular tuft to Bowman's capsule area (G/B) ratio than that of age-matched WT control while no statistically significant difference was observed at young adult age. It is worth noting that the glomerular morphology of aging *CNN2* KO mouse kidneys was mostly similar to that of WT control with ~13% having low G/B ratio. While obvious lesion only occurred in the relatively small proportion of glomeruli and did not produce clinical renal failure, severe proteinuria had developed as an early sign of renal disease.

Comparing the time course of age-progressive renal glomerular degeneration in *CNN2* KO mice with WT control, correlation analysis demonstrated that the loss of calponin 2 significantly accelerates the degeneration of glomeruli with a faster rate of age-progressive increase of low G/B ratio glomeruli (T. B. Hsieh & Jin, 2022).

6.4. *CNN2* KO Mice Develop Effacement of Podocyte Foot Processes and Increased Thickness of Glomerular Basement Membrane

Electron microscopic imaging showed that the GBM was significantly thicker in aging *CNN2* KO mouse kidneys as compared to that of age-matched WT control whereas no difference was found between young adult *CNN2* KO and WT mice. Thicker GBM is a known manifestation of kidney diseases, such as diabetic nephropathy and minimal change disease. Effacement of podocyte foot processes was found in aging but not young adult *CNN2* KO mice or young and aging WT controls. The ultrastructural data provide evidence that the loss of calponin 2 results in abnormality of podocyte structure as a functional basis for causing age-progressive glomerular degeneration and proteinuria (T. B. Hsieh & Jin, 2022).

6.5. The Role of Calponin 2 in Cytoskeleton Dynamics and Stability in Podocytes

Calponin 2 is found in a wide range of cell types originated from all three germ layers. Under normal physiological conditions, diverse types of cells express calponin 2 including cells that are under high mechanical tension such as kidney podocytes and lung alveolar cells. The expression of calponin 2 is positively regulated by mechanical tension in the cytoskeleton (Hossain et al., 2005, 2006). The cell type specific and tension-regulated expression of calponin 2 is consistent with its role in inhibiting myosin motor-driven cytoskeleton dynamics and maintaining cytoskeleton stability against the mechanical tension environment. Detachment of podocytes from the glomerular basement membrane caused by excess mechanical tension or impaired adhesion has been hypothesized as a mechanism in podocyte losses (Kriz & Lemley, 2015).

Age related loss of podocytes was reported in rodents (Bitzer & Wiggins, 2016; Hodgin et al., 2015). However, the mechanism of podocyte loss in aging kidney remains unclear. There was a strong evidence for podocyte dysfunction with loss of counterbalancing force against high hydrostatic pressure in the glomerular capillaries to cause increased GBM thickness (Butt et al., 2020). Increased GBM thickness increases the size of the filtration pores, lowering the ability of retaining albumin and other plasma macromolecules to produce proteinuria. Podocytopathies, such as that occurring in minimal change disease, focal segmental glomerular sclerosis and pre-eclampsia, share similar morphological changes in the glomerulus such as effacement of podocyte foot processes and increased GBM thickness (Craici et al., 2014). These podocytopathies develop gradually with varying degree of proteinuria, similar to the phenotype of *CNN2* KO mice.

6.6. Calponin 2 as a Molecular Factor in Podocytopathy and a Target for Treatment

The *CNN2* KO mouse studies demonstrated that calponin 2 is important for the structural and functional integrity of renal podocytes that have adapted to a unique mechanical environment due to the physiologically high hydrostatic filtration pressure in renal glomeruli. Cytoskeleton network in healthy podocytes contain noncontractile actin microfilaments in the foot processes, which are connected to the contractile actin cables in the major processes and cell body. Podocyte injury leads to the disassembly and reorganization these interconnected structure of actin framework (Suleiman et al., 2017). Calponin 2 is known to function in stabilizing noncontractile actin cytoskeleton (Panasenko & Gusev, 2001) (as that in podocyte foot processes) and to inhibit actomyosin ATPase in the contractile actin cables (Taniguchi, 2005) (as that in the cell body of podocyte). Myosin IIA plays critical roles in podocyte actin organization, contraction, and regulation of cell motility (Bondzie et al., 2016). The loss or abnormality of calponin 2 could result in dysregulated myosin motor function to increase the dynamics of actin cytoskeleton and decrease the structural stability of podocytes with decreased resistance against hydrostatic filtration-generated mechanical tension.

The finding of age-progressive proteinuria in *CNN2* KO mice supports the notion that the loss of calponin 2 regulatory function in kidney podocytes destabilizes the actin cytoskeleton, lowers the durability of podocyte foot processes, leads to the increased GBM thickness, and causes leaking glomerular filtration and proteinuria. The critical role of calponin 2 in podocyte structure and function suggests a molecular pathogenic mechanism in podocyte injury and proteinuria. A deletion of *CNN2* gene in human has been reported (Silversides et al., 2012). Therefore, genetic screening of primary proteinuria patients is needed to identify whether *CNN2* mutations contribute to the etiology. Increasing or stabilizing calponin 2 in podocytes may be explored as a therapeutic target, for which *CNN2* KO mice (Huang et al., 2008) provide an informative model system for further research and therapeutic development.

Chapter 7

Deletion of Calponin 2 Reduces Platelet Adhesion and Thrombosis

Platelet function is a critical determinant of thrombosis and hemostasis. Calponin was detected in bovine platelets and co-localized with actin during resting and stimulated conditions (Takeuchi et al., 1991). Calponin was later found in human platelets with actin colocalization in activated platelets independent of phosphorylation state (Meyer et al., 1996). Although these early studies did not identify the exact isoform of calponin expressed in platelets, the findings suggested that calponin may regulate contraction and secretion of platelets in the absence of phosphorylation.

The function of actin cytoskeleton is a major determinant of platelet activity because actin closely associates with the cytoplasmic domains of important adhesion receptors on the surface of platelets (Baig et al., 2009; Cerecedo et al., 2010; G. Gao et al., 2009; Gonzalez et al., 2012; Pertuy et al., 2014). The ability of platelets to effectively bind to collagen during physiologic flow conditions is in part dependent on the association of actin with the cytoplasmic tail of the integrin subunit through binding partners such as talin (Banno et al., 1995; Mitsios et al., 2010). To understand the role of calponin in regulating actin cytoskeleton in platelet activities may lead to new therapeutic developments for the treatments of thrombotic and hemorrhagic diseases.

7.1. Detection and Localization of Calponin 2 in Platelets

Our study then identified calponin 2 in isolated human platelets using Western blotting and immunofluorescence microscopy (Hines et al., 2014). Consistently, calponin 2 is expressed in monocytes and macrophages, which like platelets are peripheral blood cells of the myeloid lineage (Huang et al., 2008). *CNN2* KO mouse macrophages had reduced spreading area in adhesion culture together with decreased level of tropomyosin in the actin cytoskeleton when compared to calponin 2-positive macrophages (Huang et al., 2008).

Platelets have very well-described actin-mediated functions, such as spreading on adhesive surfaces, adhesion receptor localization, and clot retraction (Santos-Martínez et al., 2011). These anucleated, transcriptionally inactive fragments of megakaryocytes have a very elaborate post-translational system for regulating actin function (G. Gao et al., 2009; Neeves et al., 2008; Stalker et al., 2013), in which calponin 2 may play a role in regulating the cytoskeletal association with platelet adhesion receptors.

Immunofluorescence staining detected calponin 2 in resting human platelets and leukocytes. Calponin 2 was primarily distributed throughout the cytoplasm in resting human platelets. F-actin had a similar homogeneous distribution in resting human platelets. Activated human platelets displayed characteristic lamellipodia-like spreading on immobilized collagen and calponin 2 had a focal punctate distribution within the leading membrane edge interspersed with actin. Calponin 2 and F-actin are consistently colocalized throughout the centrally anchored region in human platelets (Hines et al., 2014).

Although platelets and erythrocytes are both from the myeloid lineage, calponin 2 was not found in peripheral mature erythrocytes. Expression of calponin 2 is known depending on the tension built in the actin cytoskeleton against the stiffness of the immobilized matrix (Hossain et al., 2005), thus the removal of this signal upon transition of erythrocyte precursor from the bone marrow to the peripheral circulation may have down-regulated calponin 2 expression and/or facilitated its degradation (Hossain et al., 2006). Platelets are fragments from parent megakaryocytes, thus more temporally associated with their nucleated progenitor, which may result in a better retention of calponin 2 in platelets compared to that in erythrocytes.

7.2. *CNN2* KO Mice Show Longer Bleeding Time and Greater Blood Loss Compared to WT Controls

Tail-bleeding assay to determine the gross differences in hemostasis phenotype of *CNN2* KO versus WT control mice demonstrated a statistically significant increase in both blood loss based on change in body weight (KO: 0.46 ± 0.08g, WT: 0.08 ± 0.03g, $p < 0.01$) and bleeding time (KO: 1005.33 ± 35.18s, WT: 437.67 ± 219.82s, $p < 0.05$). These data from a mixed arterial and venous bleeding model demonstrate that time to hemostasis is prolonged in *CNN2* KO mice (Hines et al., 2014).

7.3. *CNN2* KO Decreases Platelet Adhesion and Thrombus Formation

Physiologic shear and pulsatility significantly impact platelet activation (Neeves et al., 2008; Nesbitt et al., 2009; White et al., 2014). Therefore, a microfluidic-based whole blood thrombosis assay was employed to simulate physiologic arterial flow conditions and investigate platelet adhesion. Thrombus formation in response to collagen activation was evaluated during physiologic flow conditions to determine whether calponin 2 contributes to platelet activity with physiological relevance. Whole blood preparations from *CNN2* KO and WT mice were fluorescently labeled, flowed through the microfluidic network over immobilized collagen, and monitored by serial fluorescent photomicroscopy to document platelet and thrombus accumulation during physiological flow conditions. A significantly longer lagging time was observed for combined platelet adhesion and thrombus accumulation in *CNN2* KO mouse blood compared to WT controls (130.02 ± 3.74 s and 72.95 ± 16.23 s, respectively, $p < 0.05$) (Hines et al., 2014).

The kinetics of individual thrombus accumulation was also analyzed for whole blood from WT and *CNN2* KO mice. Despite the prolonged delaying time observed in blood from *CNN2* KO mice, the rate of thrombus accumulation was not significantly different compared to blood from WT mice (Hines et al., 2014). *CNN2* KO macrophages show faster migration and increased phagocytosis (Huang et al., 2008), reflecting higher dynamics of the actin cytoskeleton. The findings that *CNN2* KO platelets are less able to form early attachments to collagen during physiologic flow conditions due to a more dynamic actin cytoskeleton without slowing the rate of thrombus accumulation are consistent with the slower substrate adhesions of low calponin 2 prostate cancer cells (Moazzem Hossain et al., 2014) and *CNN2* KO macrophages (R. Liu & Jin, 2016b) without decreasing the maximum number of adherent cells. This feature provides a major benefit in therapeutic considerations of targeting calponin 2 for the treatment of thrombotic diseases without causing a severe hemorrhagic side effect.

7.4. Functionalities of *CNN2* KO versus WT Mouse Platelets

The subcellular distribution of calponin 2 was analyzed in mouse platelets in both resting and collagen-activated conditions. Resting platelets were fixed and permeabilized in suspension to minimize activation during the preparation. Calponin 2 was seen with a distribution throughout the cytoplasm in resting/quiescent mouse platelets with no distinct membrane association (Hines et al., 2014). This distribution is consistent with previous observations on unspecified calponin in resting bovine (Takeuchi et al., 1991) and human (Meyer et al., 1996) platelets.

Activated platelets were prepared by incubating on collagen-coated slides followed by fixation and permeabilization. Activated mouse platelets displayed characteristic spreading on immobilized collagen, and calponin 2 appeared homogenously distributed within the cytoplasmic projections and throughout the body of each platelet. F-actin appeared to have a similar homogeneous distribution throughout both resting and collagen activated platelets (Hines et al., 2014). Calponin was also shown to colocalize with actin in platelets following activation in suspension with calponin and actin translocation to the submembrane (Meyer et al., 1996; Takeuchi et al., 1991).

Calponin 2 has a focal distribution throughout the leading membrane edge lamellipodia when human and mouse platelets are activated on immobilized collagen (Hines et al., 2014). This may represent focal areas of membrane attachment to the collagen substrate. F-actin appears to be distributed more diffusely throughout the leading membrane edge, interrupted by the focal areas of calponin accumulation. Actin is known to facilitate attachment of glycoprotein VI and undergoes dynamic assembly and disassembly to facilitate membrane spreading. As in other non-muscle cell types, calponin 2 likely plays an important role in regulating this dynamic process of actin assembly and disassembly in platelets.

Platelets are unique in that calponin 2 regulation is exclusively post-translational. Although a study showed that calponin does not become phosphorylated following platelet activation (Meyer et al., 1996), it is possible that phosphorylation-dephosphorylation of calponin 2 inhibition of actomyosin ATPase activity and dynamics may regulate focal attachments required for platelet spreading on immobilized collagen. Further studies to elucidate the specific post-translational regulatory mechanisms of calponin 2 in platelets may reveal novel therapeutic targets to modify the interaction between calponin 2 and actin, and ultimately platelet functions.

7.5. Medical Significance

The finding and characterization of calponin 2 in platelets is of medical importance. Calponin 2 in platelets facilitates early interactions between platelets and collagen during physiological flow but does not significantly affect the rate or magnitude of platelet/thrombus accumulation. *CNN2* KO mice take more than double time to achieve hemostasis compared to WT controls in tail bleeding test. The kinetics of platelet adhesion and whole blood thrombosis during physiological flow conditions show that *CNN2* KO platelets are less able to form early attachments to collagen but are able to increase adhesion at a normal rate once a critical mass of adherent platelets became available to facilitate capture of flowing platelets. The data suggest that the deletion of calponin 2 slows the initiation of platelet adhesion and aggregate formation without affecting the rapid phase of platelet accumulation on collagen (Hines et al., 2014). The outcome of delaying platelet accumulation without inhibiting thrombotic potential is of significant therapeutic value. Therefore, calponin 2 may present a novel target for therapeutic platelet inhibition which can reduce pathologic thrombus initiation while minimizing spontaneous bleeding complications. Such new treatment could have significant clinical utilities, for example, in pre- and post-operational patients with an indication for platelet/thrombosis inhibition.

Chapter 8

Decreased Calponin 2 in Metastatic Cancer Cells

Tumor cells derived from various calponin-positive cell types display a common feature of decreased calponin in comparison to that in the originating tissue. For example, calponin 1 expression is significantly lower in leiomyosarcoma cells than that in normal smooth muscle cells (Horiuchi et al., 1998). Reduced calponin expressions were also found in hepatocellular carcinoma (Sasaki et al., 2002), renal angiomyolipoma (Islam et al., 2004), mammary simple carcinomas (Martín de las Mulas et al., 2004), papillary carcinomas (Mosunjac et al., 2000), basal cell-like breast carcinoma (Hasegawa et al., 2008), metastatic basal cell carcinomas (Uzquiano et al., 2008), and prostate cancer (Meehan et al., 2002; Tuxhorn et al., 2002; Verone et al., 2013). Reduced levels of calponin expression in tumor cells correspond to alterations in the dynamics and stability of actin cytoskeleton with the remaining calponin level positively related to the prognosis of the disease.

Prostate cancer is the most common malignant tumor in men in Western countries (Greenlee et al., 2000) and has a high frequency of bone metastasis (Ye et al., 2007). The molecular and cellular mechanisms leading to the development of bone metastasis of prostate cancer are not fully understood. Cancer metastasis is a process involving tumor cell proliferation, detachment, intravasation, adhesion to vascular endothelial cells, extravasation, invasion, and growth in distant organs (Fidler, 1990). Cell motility plays a fundamental role in every step of this process, and the function of actin cytoskeleton is vital for cell division (E. K. H. Han et al., 1993) and migration (Ponti et al., 2004).

Alterations of actin (Okamoto-Inoue et al., 1999; Shimokawa-Kuroki et al., 1994) and actin-associated proteins, including calponin 1 (Horiuchi et al., 1998), have been implicated in cancer metastasis. Studies have found a loss of calponin 1 in prostate cancer in comparison with non-cancerous prostate tissues (Meehan et al., 2002; Moazzem Hossain et al., 2014). Calponin 1 is known to express specifically in smooth muscle cells (R. Liu & Jin, 2016a) (see Figure 3A in Chapter 1). Therefore, its presence in prostate tissue and decrease in prostate cancer may reflect the expression in stroma smooth muscle-like cells (Chakrabarty et al., 2019). The decreased level of calponin 1 in prostate cancer may therefore indicate a destruction of smooth muscle-like stroma cells (Gerdes et al., 2004) whereas the function of calponin 2 in the non-muscle actin cytoskeleton of prostate cancer cells may contribute to the metastatic properties.

8.1. Presence of Calponin 2 in Prostate Epithelial Cells with Decreases in Prostate Cancer

Calponin 2 is expressed in smooth muscles (Strasser et al., 1993) as well as in various non-muscle cells, including epithelial cells (Fukui et al., 1997; Hossain et al., 2005, 2006) (see Figure 3B in Chapter 1). Western blots of total protein extract detected calponin 2 in non-cancerous human prostate tissue, which was localized to the epithelial cells by immunohistochemistry stain of the prostate gland and diminished in prostate cancer tissues (Moazzem Hossain et al., 2014). Calponin 1 was not detected in prostate epithelial cells. A comparison between a metastatic human prostate cancer cell line PC3-M and its parental cell line PC3 (Kozlowski et al., 1984) showed that normalized to the level of actin, densitometry quantification of Western blots found that the level of calponin 2 in PC3-M cells was ~34% lower than that in PC3. PC3-M cells demonstrated faster rates of proliferation and migration than that of PC3, implicating a negative correlation of calponin 2 level to the metastatic phenotype. Similarly, cultured primary human cancerous prostate cells had a significantly lower level of calponin 2 than that in non-cancerous controls (Moazzem Hossain et al., 2014). Confocal microscopic images of PC3 and PC3-M cells showed that calponin 2 co-localized with F-actin, consistent with the function of calponin 2 in actin-mediated cellular functions such as migration and proliferation (Hossain et al., 2003; Huang et al., 2008). Fluorescence intensity of the confocal images confirmed that PC3-M cells contained a lower level of calponin 2 than that in PC3. There was no qualitative difference in the cellular distribution of calponin 2 in PC3 and PC3-M cells (Moazzem Hossain et al., 2014).

Another study also found calponin 2 in prostate cancer cells and loss of calponin 2 induced cellular protrusions and increased cell migration (Verone et al., 2013). It is worth noting that decreased calponin 2 was not uniformly seen in all prostate cancers examined, but with a strong correlation to cell motility features. Therefore, decreases in calponin 2 may be used as a potential biomarker for evaluating the degree of malignancy of prostate cancer in individual patients.

8.2. Loss of Calponin 2 in Prostate Cancer Cells Increases Cell Proliferation

Correlating to the lower level of calponin 2, the growth curve of PC3-M cells in culture showed a significantly faster rate of proliferation than that of PC3 cells. The inverted correlation between the level of calponin 2 and the rate of tumor cell proliferation was confirmed by the faster proliferation rate of immortalized primary prostate cancer cells expressing low versus high levels of calponin 2 (Moazzem Hossain et al., 2014). To demonstrate the causal effect of decreased calponin 2 on the increased proliferation of prostate cancer cells, boosting the level of calponin 2 in PC3-M cells by stable transfective expression effectively suppressed the rate of proliferation (Moazzem Hossain et al., 2014). Supporting the tumor suppressor function of calponin in inhibiting malignancy of tumor cells, transfective expression of calponin 1 in human fibrosarcoma, leiomyosarcoma, synovial sarcoma and osteosarcoma cells also significantly reduced anchorage-independent growth and in vivo tumorigenecity (Horiuchi et al., 1999; Takeoka et al., 2002; Yamamura et al., 2001). A more recent study showed that calponin 2 is an inhibitory regulator for the rate of cytokinesis (Qian et al., 2021), demonstrating a mechanism underlying its role in inhibiting tumor cell proliferation.

8.3. Loss of Calponin 2 in Prostate Cancer Cells Increases Cell Migration

The lower level of calponin 2 in PC3-M cells in comparison to that in PC3 cells also corresponds to a faster rate of cell migration. In vitro scratch wound healing assay showed that PC3-M cells healed the wound in monolayer culture faster than that of PC3 (Moazzem Hossain et al., 2014), consistent with the observation of calponin 2 inhibition of macrophage migration (Huang et al., 2008; R. Liu & Jin, 2016b). Therefore, the decreased level of calponin 2 in PC3-M cells may contribute to its high motility and metastatic phenotype in comparison to its parental PC3 cells (Kozlowski et al., 1984). To demonstrate that the effect of decreased calponin 2 on the motility of prostate cancer cells is not restricted to the PC3-M versus PC3 cell lines, the correlation between decreases in calponin 2 and increased rate of cell migration was further demonstrated in non-malignant versus malignant/cancerous prostate endothelial cells (Moazzem Hossain et al., 2014). The effect of low calponin 2 on the faster rate of PC3-M versus PC3 cell migration in wound healing assays was further shown in cells with high and low levels of stable transfective expressions calponin 2 (Moazzem Hossain et al., 2014). The results support a causal role of decreased calponin 2 in the high motility phenotype of metastatic prostate cancer cells.

The universal inhibitory effect of calponin 2 on cell proliferation and migration was also shown by the significantly faster proliferation and migration rates of macrophages from *CNN2* KO mice in comparison to that of WT mouse macrophages expressing a high level of calponin 2 (Hossain et al., 2003; Huang et al., 2008; R. Liu & Jin, 2016b). These data suggest that decreases in calponin 2 in prostate cancer cells may contribute to malignant growth and metastasis through facilitating remodeling of the actin cytoskeleton during cell proliferation and migration.

8.4. Loss of Calponin 2 Decreases Substrate Adhesion of Prostate Cancer Cells

The decrease of calponin 2 in PC3-M cells results in weakened substrate adhesion as compared with that of PC3 cells. More PC3 cells than PC3-M cells were attached to uncoated plastic cell culture plates in the first 8 hours after plating. This difference was further shown in adhesion experiments using collagen-coated culture plates. ~53% of PC3 cells attached 10 minutes after seeding versus ~20% for PC3-M cells ($p < 0.001$). Such effect of calponin 2 on cell adhesion was also seen in normal versus cancerous primary prostate cells expressing high and low calponin 2, respectively (Moazzem Hossain et al., 2014).

The proportion of adhered cells became similar between PC3 and PC3-M 24 hours after seeding to uncoated plates and 1 hour after seeding to collagen-coated plates (Moazzem Hossain et al., 2014). This preserved maximum adhesion suggests that calponin 2 functions in accelerating the speed, other than the capacity, of cell adhesion. The similar trends of adhesion of the cells to uncoated and collagen-coated plates indicate that the effect of calponin 2 on cell adhesion is independent of collagen receptors, in which the slower adhesion to uncoated plates reflected the need of cell-secreted collagen to facilitate adhesion.

Supporting the role of calponin 2 in enhancing cell adhesion, the spreading area of PC3-M cells on plastic substrate was significantly smaller than that of PC3 cells. Consistent with the role of calponin 2 in stabilizing actin cytoskeleton (Hossain et al., 2005) as a mechanism in enhancing cell adhesion, the lower level of calponin 2 in PC3-M cell correlated to faster roundup during trypsin digestion of adherent monolayer cultures than that of PC3 cells expressing higher level of calponin 2. Transfective expression to boost calponin 2 in PC3-M cells effectively slowed the roundup velocity (Moazzem Hossain et al., 2014).

8.5. Loss of Calponin 2 Increases the Dependence of Prostate Cancer Cell Adhesion on Substrate Stiffness

It is widely observed that cells adhere faster to culture substrate of higher stiffness. An interesting finding was that the low calponin 2-produced weaker adhesion of PC3-M cells is accompanied by an increased dependence on substrate stiffness (Moazzem Hossain et al., 2014). Shown in Figure 5, collagen-coated high stiffness polyacrylamide gel substrate had significantly greater enhancement for the adhesion of low calponin 2 PC3-M and primary prostate cancer cells than that of control cells expressing higher levels of calponin 2. The data suggest that the decreased level of calponin 2 in prostate cancer cells increases the preference of adhesion to high stiffness substrates. This mechanism may contribute to prostate cancer's high tendency of metastasis to the bone that provides a uniquely high stiffness tissue substrate.

Comparing high versus low calponin 2 expressing prostate cells and wild type versus calponin 2 KO mouse fibroblasts for their adhesions to hard and soft gel substrates, the results showed that all these cell lines adhered with lower numbers to low stiffness soft substrate than that to high stiffness hard substrate, similar to that of PC3 versus PC3-M. In contrast, there was no difference between a control pair of prostate cells that express similar levels of calponin 2 (Moazzem Hossain et al., 2014).

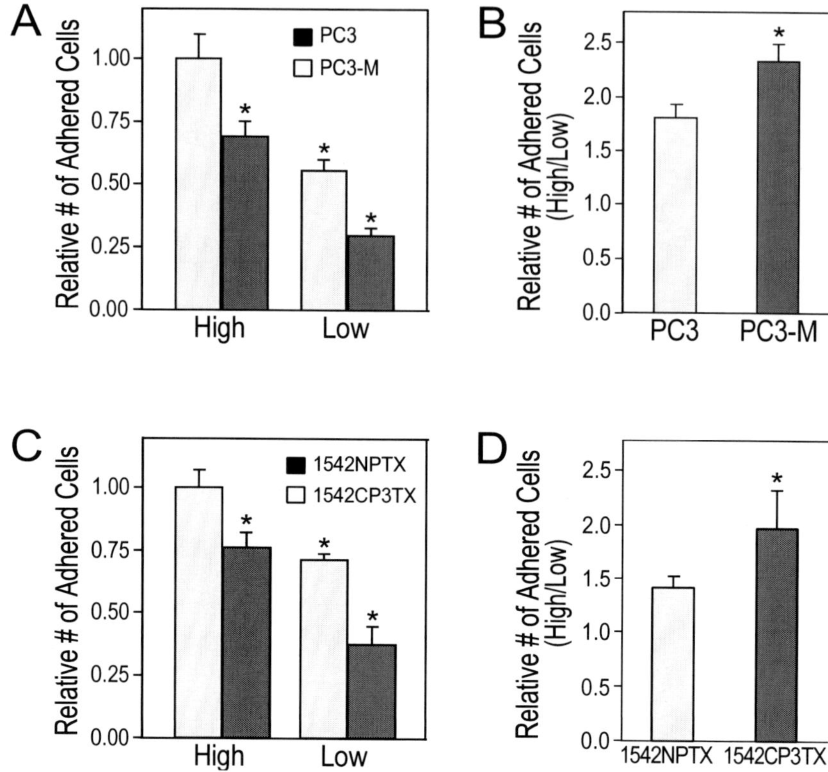

Figure 5. Decreased calponin 2 in prostate cancer cells increases the dependence of cell adhesion on substrate stiffness. Demonstrating the negative correlation between calponin 2 level and substrate stiffness dependent cell adhesion, the results of cell adhesion to polyacrylamide gel substrates showed that the low calponin 2 PC3-M (A) and 1542CP3TX (C) prostate cancer cells adhered less in comparison with that of PC3 and 1542NPTX cells expressing higher levels of calponin 2. Data analysis for the ratio of cells adhered to high versus low stiffness gel substrates (B and D) further showed that cells expressing lower level of calponin 2 had significantly higher increases in adhesion to high stiffness substrate versus low stiffness substrate than that of cells expressing higher levels of calponin 2. Data are presented as mean ± SD, $*p < 0.001$.

8.6. Substrate Stiffness Determines the Expression of Calponin 2 in Prostate Cancer Cells

The expression of calponin 2 is regulated by the stiffness of culture substrate (Hossain et al., 2005). In comparison to the effect of nearly rigid plastic or high stiffness gel substrates, soft gel substrates that produce lower traction force in the cytoskeleton (Polte et al., 2004) decreased calponin 2 expression in epidermal keratinocytes, fibroblasts (Hossain et al., 2005) and lung alveolar cells (Hossain et al., 2006). This is also true for PC3 and PC3-M prostate cancer cells. PC3 and PC3-M cells cultured on low stiffness soft polyacrylamide gel (elastic modulus = 1 kPa) expressed significantly less calponin 2 than that in cells cultured on higher stiffness hard gel (elastic modulus = 8 kPa). The high-level expression of calponin 2 in cells grown on very hard gel substrate (elastic modulus = 75 kPa) was similar to that in cells grown on plastic dish, indicating that the physical stiffness, other than the chemical property, of the cultural substrates determined the level of calponin 2 in prostate cancer cells (Hossain et al., 2005). Together with studies in other cell types, the results demonstrate a conserved regulation of calponin 2 gene expression by cytoskeleton tension in response to the stiffness of culture substrate. The high stiffness environment will increase the expression of calponin 2 in adhered cells as positive feedback to stabilize cell adhesion and facilitate the bone metastasis of prostate cancer cells, a mechanism especially important for cancer cells that had primarily decreased expression of calponin 2 to maintain a stable actin cytoskeleton and settle in the bone environment.

8.7. Medical Implications

Extensive studies have investigated the role of the chemical environment of bone tissue in prostate cancer's high tendency of bone metastasis (Ye et al., 2007). It has been observed that living cells also dynamically change structure and function through gene regulation and posttranslational protein modification in response to mechanical environment (Chicurel et al., 1998; Eckes & Krieg, 2004; Hamill & Martinac, 2001; Lehoux & Tedgui, 2003; J. L. Walker et al., 2005). The potential contribution of the high stiffness mechanical environment in bone tissue to the metastasis of prostate cancer presents a new direction of research. Based on calponin 2's established role as a cytoskeleton regulatory protein (J.-P. Jin et al., 2008; R. Liu & Jin, 2016a; K. G. Morgan & Gangopadhyay, 2001) which is responsive to the mechanical tension environment (Hossain et al., 2005, 2006; Jiang et al., 2014; Rasmussen & Jin, 2024), the metastatic potency of low calponin 2 prostate cancer cells correlating to the increased dependence of cell adhesion on substrate stiffness indicates a novel pathological function of calponin 2 mechanoregulation.

The actin cytoskeleton undergoes dynamic remodeling during cell proliferation, adhesion and migration, which are key steps in tumor progression and metastasis. By affecting the structural organization (Danninger & Gimona, 2000; Fukui et al., 1997) and stability (Hossain et al., 2005) of actin cytoskeleton, reducing the rates of cell proliferation by inhibiting cytokinesis (Hossain et al., 2003; Qian et al., 2021) and cell migration, and facilitating substrate adhesion (Moazzem Hossain et al., 2014), calponin 2 may present an attractive molecular target for the treatment of cancer metastasis especially for the reduction of bone metastasis.

Chapter 9

Deletion of Calponin 2 in Macrophages Attenuates Inflammatory Arthritis

Rheumatoid arthritis is a multifactorial autoimmune disease characterized by chronic joint inflammation, which leads to progressive joint destruction and ultimately to significant disability and severely reduced quality of life. Macrophage-mediated inflammation is a critical part of the pathogenesis of inflammatory diseases such as systemic lupus erythematosus and rheumatoid arthritis (Kinne et al., 2000; Y. G. Lee et al., 2011). Calponin 2 is expressed in myeloid leukocytes, i.e., granulocytes, monocytes and macrophages and functions as an inhibitory regulator of cell proliferation, macrophage migration and phagocytosis (Huang et al., 2008). Macrophages are mobile cells, and their actin cytoskeleton plays a vital role in motility-based activities such as migration, adhesion and phagocytosis. The phagocytotic function of macrophages is essential for clearance of necrotic and apoptotic cells, pathogens and other harmful substances, presenting a therapeutic mechanism for promoting the resolution of inflammation (Hallett et al., 2008; Y. G. Lee et al., 2011; G. Liu et al., 2006; Silva, 2011). Defective macrophage phagocytosis has been related to the pathogenesis of systemic lupus erythematosus (Y. Li et al., 2010) and rheumatoid arthritis (Paino et al., 2011; Tas et al., 2006). The functions of macrophage cytoskeleton are, therefore, essential for the activities of macrophages in health and in the development and resolution of inflammation (Y. G. Lee et al., 2011), in which calponin 2 plays a regulatory role (Huang et al., 2016).

9.1. Up-Regulation of Calponin 2 in Macrophages in the Joints of Rheumatoid Arthritis Patients and Mouse Model

Immunohistochemistry staining of paraffin sections of joint synovial tissues from rheumatoid arthritis patients using an anti-calponin 2 mAb 1D2 detected significantly higher levels of calponin 2 in more cells than that in samples from arthritis-free control subjects. The calponin 2-positive cells in the joints of rheumatoid arthritis patients were mainly in the lining and sublining of synovial tissue and morphologically macrophage-like (Huang et al., 2016).

To assess the nature of the cells with increased calponin 2, purified peripheral blood $CD14^+$ monocytes and synovial fluid macrophages from rheumatoid arthritis patients were examined in Western blot using mAb 1D2 in comparison with monocytes from healthy donors and the in vitro differentiated macrophages. Significantly higher levels of calponin 2 were found in peripheral blood monocytes and joint fluid macrophages of rheumatoid arthritis patients, suggesting a possibly systemic increase of calponin 2 expression in these cells in inflammatory disease. The findings implicate that increased levels of calponin 2 in macrophages may contribute to the pathogenesis of inflammatory arthritis (Huang et al., 2016). It remains to be investigated whether the higher level of calponin 2 is a precondition for increased susceptibility to the development of, or an initial change in, the pathogenesis of inflammatory diseases such as rheumatoid arthritis.

Anti-glucose-6-phosphate isomerase (GPI) antibodies are found in many rheumatoid arthritis patients in association with more severe forms of disease (Maccioni et al., 2002). In a mouse arthritis model, Western blots and densitometry quantification showed that calponin 2 increased significantly in tissue homogenates of ankle joints during the course of anti-GPI serum transfer-induced arthritis (Monach et al., 2008, 2010) in concurrence with the development of arthritic inflammation (Huang et al., 2016). The level of calponin 2 in the mouse ankle homogenates was positively correlated with the degree of joint inflammation with statistical significance. This correlation supports the observation that the high levels of calponin 2 in synovial macrophages of rheumatoid arthritis patients may contribute to the development of pathological inflammation.

Synovial fibroblasts also play a role in the development of inflammatory arthritis (Yamamura et al., 2000). The calponin 2-positive cells increased in the synovial lining of rheumatoid arthritis patient joints (Huang et al., 2016) may include both macrophages and fibroblasts. Therefore, synovial fibroblasts may also have increased calponin 2 in rheumatoid arthritis.

9.2. Myeloid Cell-Specific Deletion of Calponin 2 Attenuated the Severity of Anti-GPI Serum-Induced Arthritis

The effect of calponin 2 deletion on the pathogenesis and pathology of inflammatory arthritis was investigated using the anti-GPI serum-transfer mouse model. Compared with WT littermates, systemic *CNN2* KO mice showed significantly milder joint inflammation at days 0, 2, 4, 7 and 9 post arthritis induction with less hind ankle swelling (smaller circumference) and lower grades of clinical index for all four paws. Quantitative analysis of H&E-stained ankle joint sections collected on day 9 after the induction of arthritis demonstrated significantly less intra- and extra-articular inflammatory cell infiltration and bone erosion in *CNN2* KO mice than that in WT control. The results suggest that calponin 2 regulation of cytoskeleton function plays a novel role in the development of inflammatory arthritis (Huang et al., 2016).

Mice with a conditional calponin 2 deletion specifically in myeloid cells were then studied to determine whether calponin 2-dependent functions of macrophage contribute to the pathogenesis of inflammatory arthritis. $CNN2^{f/f}$,$lysM^{cre}$ mice were generated by crossing *CNN2*-floxed mice (Huang et al., 2008) with $lysM^{cre}$ mice. $lysM^{cre}$ mice carry a transgene encoding Cre recombinase under the myeloid cell-specific *lys*M promoter that is expressed in granulocytes, monocytes, macrophages, and osteoclasts but not T and B lymphocytes, dendritic cells (Clausen et al., 1999; Huang et al., 2010) or fibroblasts. In addition to PCR genotyping, the deletion of calponin 2 in macrophages was verified using Western blots of protein extracts from thioglycolate-elicited peritoneal macrophages. The result showed that calponin 2 became undetectable in $CNN2^{f/f}$,$lysM^{cre}$ mouse macrophages in contrast to the significant expression in macrophages elicited from $CNN2^{f/f}$ mice with the expression unaffected in the skin (a representative control tissue) (Huang et al., 2016).

Confirming the phenotype of systemic *CNN2* KO mice, myeloid cell-specific *CNN2* KO very effectively attenuated the severity of anti-GPI serum-transfer-induced arthritis as shown by the significantly lower clinical arthritis scores and less ankle swelling than that of WT control. Very little calponin 2 is normally expressed in mature granulocytes. Therefore, the attenuated development of arthritis in myeloid cell-specific *CNN2* KO mice strongly supports that calponin 2-modulated macrophage functions make a leading contribution to the pathogenesis of inflammatory arthritis (Huang et al., 2016).

9.3. Calponin 2 Regulates Motility Related Functions of Macrophage in Inflammatory Arthritis

Actin cytoskeleton-based cell motility plays an essential role in the functions of macrophages (Calle et al., 2006). Based on its inhibitory regulation of the actin-activated myosin motors (Hossain et al., 2016; R. Liu & Jin, 2016a), deletion of calponin 2 and the inhibition of myosin activity increases macrophage migration and phagocytosis (Huang et al., 2008). The migration velocity of calponin 2-null versus WT mouse macrophages in the absence of chemotactic stimulation examined using in vitro wound healing assay and time lapse microscopy found that the deletion of calponin 2 increased migration velocity of macrophages (Huang et al., 2016). This observation is consistent with the effect of calponin 2 deletion in other cell types (Moazzem Hossain et al., 2014), reflecting a calponin regulation of cytoskeleton dynamics.

The correlation between the increase of calponin 2 in macrophages and the development of inflammatory arthritis and the dramatic effect of calponin 2 deletion on attenuating the severity of inflammatory disease suggests a hypothesis that the increased level of calponin 2 in macrophages of rheumatoid arthritis patients may reduce cell motility and phagocytotic activities which are essential for trafficking towards the affected site and the clearance of cellular debris, apoptotic cells and inflammatory products. A high level of calponin 2 would result in reduced scavenging of inflammatory stimuli and hindering the resolution of inflammation, whereas the deletion of calponin 2 in macrophages enhances these functions to facilitate the resolution of arthritic inflammation. Increased phagocytosis has also been implicated for a role in promoting macrophage differentiation toward an anti-inflammatory phenotype (Galli et al., 2011). suggesting another mechanism for the deletion of calponin 2 to attenuate arthritic inflammation.

9.4. *Cnn2* KO Does Not Alter the Baseline Immunophenotype of Mouse Spleen and Peripheral Blood Cells

Flow cytometry immunophenotyping profiles of CD11b+ myeloid splenocytes from *CNN2* KO and WT mice, including granulocytes ($CD64^-$, $Ly6G^+$), tissue residential macrophages ($CD64^+$, $F4/80^{hi}$, $Ly6c^-$) and monocyte-derived macrophages ($CD64^+$, $F4/80^{lo}$, $Ly6c^+$), were compared. Their peripheral neutrophils, lymphocytes and monocytes were analyzed by blood count, with CD11b^{+}, CD115^{+} monocytes further analyzed by flow cytometry for the subsets of classic monocytes (Ly6c^{+}, CD62L^{+}) and non-classic monocytes (Ly6c^{-}, CD62L^{-}). The results showed that deletion of calponin 2 did not cause detectable change in the baseline phenotype of myeloid cells, except for a slight increase of F4/80^{lo}, CD11b^{+}, Ly6C^{+} macrophages (Huang et al., 2016). While no change was observed in the populations of monocytes, this phenotype may be an implication for a role of calponin 2 in the differentiation of circulation monocytes into tissue macrophages.

9.5. Role of Calponin 2-Regulated Cell Adhesion in Macrophage Activation and Differentiation

Cell adhesion is a critical factor in macrophage differentiation and activation (Szekanecz & Koch, 2007; H. Liu et al., 2008). Interaction of monocytes with extracellular matrix affects their differentiation into macrophages (Wesley et al., 1998). Monocyte-macrophage differentiation is accompanied by integrin/CD11b expression and cell adhesion. Treatment with CD11b antisense RNA reduced monocyte adhesion and attenuated their differentiation into macrophages (Prudovsky et al., 2002). The deletion of calponin 2 caused slower adhesion of macrophages to cultural dishes. The maximum number of adherent cells at 24 hours was, however, not affected (Huang et al., 2016). The effect of calponin 2 deletion on decreasing the adhesion velocity of macrophages is consistent with its similar effects in the studies on the adhesion of prostate cancer cells (Moazzem Hossain et al., 2014) and platelets (Hines et al., 2014).

The effect of calponin 2 deletion on decreasing the velocity of macrophage adhesion suggests a hypothesis that the increased calponin 2 in macrophages from rheumatoid arthritis patients would facilitate cell adhesion and promote pathological activation and differentiation into proinflammatory phenotypes. Supporting this hypothesis, deletion of calponin 2 in macrophages resulted in effective attenuation of arthritis as well as increased phagocytosis and rate of proliferation (Huang et al., 2016), both are anti-inflammatory phenotypes (Stables et al., 2011; Weigert et al., 2006).

9.6. Deletion of Calponin 2 Inhibits Osteoclastogenesis and Bone Resorption Activity of Osteoclasts

In the anti-GPI serum transfer-induced arthritis model, calponin 2-null mice developed milder joint inflammation with significantly less bone erosion in comparison with that in WT mice (Huang et al., 2016). Osteoclasts are derived from myeloid cell lineage and are effector cells for bone erosion in arthritis joints (Boyce et al., 2006). Demonstrating that deletion of calponin 2 decreases osteoclastogenesis and the bone resorption activity of osteoclasts, *CNN2* KO mouse bone marrow cells containing osteoclast precursors (OCPs) formed significantly fewer osteoclasts upon M-CSF and RANKL-simulation in culture than that from cells of WT littermates. The osteoclasts derived from *CNN2* KO mouse bone marrow cells produced fewer resorption pits on bone slices than that produced by WT osteoclasts (Huang et al., 2016).

Both $CD45^+c\text{-}Kit^+c\text{-}Fms^+CD11b^+$ and $CD45^+c\text{-}Kit^+c\text{-}Fms^+CD11b^-$ cells give rise to osteoclasts and represent bone marrow OCPs (P. Li et al., 2004). The decreased osteoclast formation in *CNN2* KO mice is associated with reduced OCPs. The frequency of OCPs in bone marrow cells by flow cytometry was slightly decreased in the *CNN2* KO group. Bone morphometric analysis on hind limb sections demonstrated that bone volume and osteoclast numbers were similar in 3-month-old *CNN2* KO and WT mice, indicating that deletion of calponin 2 did not significantly affect basal osteoclast formation and function (Huang et al., 2016). However, calponin 2 may be required for stimulated osteoclast generation and activation, both are critical factors in the joint destructions in rheumatoid arthritis.

An established function of calponin 2 is the stabilization of actin cytoskeleton via the inhibition of myosin motors (Hossain et al., 2005), which may be a mechanism for its inhibitory effects on cell proliferation, migration, adhesion and other motility-related activities. Osteoclast-mediated bone erosion in arthritic joints is based on adhesion and the formation of an actin ring (Harre et al., 2012). Therefore, deletion of calponin 2 may have a direct effect on reducing the ability of osteoclasts to form the actin ring and resorption pits, thus attenuating bone erosion as seen in the anti-GPI serum-induced mouse arthritis (Huang et al., 2016). The expression of calponin 2 is cell adhesion-dependent (Hossain et al., 2005; Jiang et al., 2014) and calponin 2 enhances cell adhesion to the extracellular substrate. The high level of calponin 2 in macrophages during inflammation may, therefore, increase the accumulation of macrophages in the joint and promote macrophage activation and osteoclast formation.

9.7. Calponin 2 as a Novel Cytoskeleton Regulator of Macrophage Function in Inflammatory Arthritis and Potentially a Therapeutic Target

In addition to the functions of calponin 2 in regulating macrophage migration and phagocytosis (Huang et al., 2008), the calponin 2-dependent cell adhesion may be critical to the differentiation and activation of macrophages, as well as the bone resorption activity of osteoclasts (Tanaka et al., 2005). Therefore, the function of calponin 2 in regulating interactions between macrophages and the tissue mechanical environment (Rasmussen & Jin, 2024) may play a novel role in the pathogenesis of inflammatory arthritis.

With the finding that macrophages in the joints of rheumatoid arthritis patients had elevated levels of calponin 2, the role of calponin 2 in regulating cytoskeleton function in the pathogenesis of inflammatory disease such as arthritis implicates a potentially therapeutic target. Many of the current therapies for rheumatoid arthritis target upstream receptors and pathways in the signaling hierarchy of inflammatory responses, which, however, have broad effects on multiple cell types and may cause multiple functional changes in affected cells including macrophages. In contrast, the calponin 2 regulation of cytoskeleton and cell motility-based functions of macrophages demonstrates a downstream effector mechanism with relatively restrictive consequences. Supporting the hypothesis that deletion of calponin 2 in macrophages is a downstream cellular mechanism to attenuate inflammatory arthritis, the levels of representative pro- and anti-inflammatory cytokines IL-1β, TNFα and IL-10 in the tissue extracts of the inflammatory joints of *CNN2* KO mice were similar to that in WT controls (Huang et al., 2016). Therefore, targeting the cytoskeleton-based function of macrophages through reducing the expression or facilitating the degradation of calponin 2 may steer macrophage function toward the resolution of inflammation for the treatment of inflammatory arthritis without altering the systemic pro-inflammatory environment.

Chapter 10

Deletion of Calponin 2 in Macrophages Minimizes the Development of Hyperlipidemia-Induced Arterial Atherosclerosis

Arterial atherosclerosis is an inflammatory disease (Libby, 2012; Nahrendorf & Swirski, 2015; Ross, 1999) and the primary cause of ischemic heart disease and stroke (Libby, 2012). Increasing levels of circulating low-density lipoprotein (LDL)-cholesterol and the subsequent intramural accumulation of oxidized LDL trigger the recruitment and retention of monocytes to generate subendothelial lesions in arterial wall (Lusis, 2000). In the intima of blood vessel wall, monocytes differentiate into macrophages to scavenge lipoprotein particles and become foam cells, which is a landmark of atherosclerosis (Lusis, 2000). Macrophages and the lipid ingestion-generated foam cells play critical roles in mediating the ensuing inflammatory response and prognosis of atherosclerosis plaques (Chinetti-Gbaguidi et al., 2015) in the pathogenesis and pathology of atherosclerosis (Dickhout et al., 2008). The regulation of macrophage activation and function is at the center of this process and a target for therapeutic development.

Differentiation and phenotype polarization of macrophages promote either resolution of the inflammatory process and attenuation of atherogenesis (Cardilo-Reis et al., 2012; N. Sharma et al., 2012) or acceleration of atherosclerosis (Hamers et al., 2012; Hanna et al., 2012). The motility and substrate adhesion of macrophages are essential in the development and resolution of inflammation (Patel et al., 2012; Sosale et al., 2015). To investigate the role of calponin 2 that regulates the functions of macrophages (Huang et al., 2008) in the pathogenesis of atherosclerosis, the effect of deleting calponin 2 in macrophages on the development of hyperlipidemia-induced arterial atherosclerosis was studied using cellular and in vivo mouse models (R. Liu & Jin, 2016b).

10.1. Deletion of Calponin 2 in Macrophages Attenuated the Development of Aortic Atherosclerosis in *Apo*E$^{-/-}$ Mice

Apolipoprotein E (*Apo*E) plays an important role in cholesterol transportation and metabolism. *Apo*E deficiency in mice leads to hyperlipidemia and spontaneous atherosclerosis even when fed with a normal diet (Daugherty, 2002). To study the effect of calponin 2 deletion on the function of macrophages in the pathogenesis of atherosclerosis under a physiologically relevant condition, *Apo*E$^{-/-}$ mice fed on normal chow diet were used as an atherosclerosis model. *Apo*E$^{-/-}$; *Apo*E$^{-/-}$,*CNN2*$^{-/-}$ double KO; and *ApoE*$^{-/-}$,*CNN2*$^{f/f}$,*lysM*$^{cre+}$ myeloid cell-specific *CNN2* KO mice were studied at 6.5 months of age to examine aortic atherosclerotic lesions. A possible gender difference in the development of atherosclerosis in *Apo*E$^{-/-}$ mice has been suggested (Chiba et al., 2011; Coleman et al., 2006; Meyrelles et al., 2011; Tangirala et al., 1995). Therefore, data of both male and female mice were collected to consider gender-based variations. The aorta en face staining method was used to quantify atherosclerotic lesions.

While no significant difference was detected in the levels of total serum cholesterol between *ApoE*$^{-/-}$ and *ApoE*$^{-/-}$,*CNN2*$^{-/-}$ mice (537±157 mg/dL and 519±176 mg/dL, respectively, mean ± SEM), in male mice the area of *Apo*E deficiency-caused atherosclerotic plaques was reduced by 56.2% with systemic deletion of calponin 2 ($p < 0.05$) and by 91.9% with myeloid cell-specific deletion of calponin 2 ($p < 0.01$). In females, *ApoE*$^{-/-}$,*CNN2*$^{-/-}$ and *ApoE*$^{-/-}$,*CNN2*$^{f/f}$,*lysM*$^{cre+}$ mice also showed significantly reduced plaque areas (51.9%, $p < 0.01$ and 50.3% less as compared with *ApoE*$^{-/-}$ controls, respectively, $p <$ 0.0.1). The total area and macrophage content of atherosclerotic lesions at aortic roots of male *ApoE*$^{-/-}$,*Cnn2*$^{f/f}$,*lysM*$^{cre+}$ mice were significantly less than that of the control *ApoE*$^{-/-}$ mice (R. Liu & Jin, 2016b).

While estrogen plays an atheroprotective role in female mice (Thomas & Smart, 2007), there are data indicating that the lesions in *ApoE*$^{-/-}$ mice were more severe in females (Caligiuri et al., 1999; Meyrelles et al., 2011). Consistently, the *CNN2* KO study showed more severe lesion development in female *ApoE*$^{-/-}$ mice and a more effective attenuation of atherosclerosis by calponin 2 deletion in male mice (R. Liu & Jin, 2016b).

The results showed that myeloid cell-specific KO of *CNN2* has the same or stronger effect in comparison with that of systemic KO, indicating that the therapeutic effect was primarily via the function of myeloid cells. This observation is consistent with the finding that myeloid cell-specific *CNN2* KO had stronger effects on attenuating inflammatory arthritis than that of global *CNN2* KO (Huang et al., 2016). These data also implicate that the loss of calponin 2 in some other cell types may counteract the effect of calponin 2 deletion in myeloid cells on the attenuation and resolution of inflammations.

LysM-cre induces *CNN2* KO in macrophages as well as in other myeloid cells. *CNN2* KO did not alter the subsets of Ly6C^{hi} and Ly6C^{low} monocytes, which are considered the sources of inflammatory and anti-inflammatory macrophages, respectively. No subset difference was detected in peritoneal residential macrophages from WT and *CNN2*$^{-/-}$ mice (Huang et al., 2016). These data support that *CNN2* KO in myeloid cells attenuates the pathogenesis of atherosclerosis via post-monocyte mechanisms (R. Liu & Jin, 2016b).

10.2. *CNN2*$^{-/-}$ Macrophages Retain the Ability of Lipid Uptake

The distribution patterns of F-actin and myosin are similar in WT versus *CNN2*$^{-/-}$ macrophages and foam cells with F-actin concentrated in the leading edge and the trailing tail of migrating macrophages whereas myosin mainly in the center of cell body. The cellular location of calponin 2 in WT macrophages and foam cells is similar to that of myosin IIA (R. Liu & Jin, 2016b). A possibly underlying mechanism is that calponin 2 enhances substrate adhesion of cells by inhibiting myosin motor activity and decreasing the dynamics of cytoskeleton (Hossain et al., 2016).

CNN2$^{-/-}$ macrophages showed an enhanced phagocytotic activity when assessed by the uptake of mouse serum-coated fluorescent latex beads (Huang et al., 2008). Different from the non-specific phagocytosis of beads, the uptake of lipid complex by macrophages is a scavenger receptor-mediated engulfment which is also related to the intracellular cholesterol metabolism in macrophages. Peritoneal residential macrophages from *CNN2*$^{-/-}$ and WT mice were examined for the effect of calponin 2 deletion on the lipid uptake ability of macrophages and the formation of foam cells after 4, 8 and 24 hours of incubation with acetylated LDL. The results showed that intracellular lipid droplets were significantly increased during the course of incubation similarly in both WT and *CNN2*$^{-/-}$ groups (R. Liu & Jin, 2016b). The retained lipid uptake ability of *CNN2*$^{-/-}$ macrophages indicates that the effect of *CNN2* KO on attenuating atherosclerosis is not via a reduction of lipid engulfment.

10.3. Deletion of Calponin 2 Alleviates the Impaired Motility of Lipid-Laden Foam Cells

Macrophages and foam cells are pivotal cell types in the development of inflammatory lesion in arterial atherosclerosis, effecting on the progression and regression of plaques (Moore et al., 2013). Although phagocytosis clearance of lipoproteins by macrophages is likely to be beneficial at the outset of this inflammatory response, dysregulation of lipid metabolism and accumulation of lipid-ingested macrophages in atherosclerotic plaques may alter immune phenotypes and cause apoptosis to exaggerate inflammatory response and aggravate the progression of atherosclerosis lesion. Therefore, promotions of cholesterol efflux from macrophages (Ohashi et al., 2005) and macrophage emigration from plaques (van Gils et al., 2012) have been proposed as therapeutic approaches.

Cholesterol loading in macrophages results in significant reduction of migration ability (Pataki et al., 1992; Qin et al., 2006), accompanied by a decreased capacity of force generation by cell locomotors (Zerbinatti & Gore, 2003). Lipid-ingested macrophages have hindered migration and the retention of macrophages in atherosclerotic lesions contributes to the failure of resolving inflammation and the development of plaques (Ludewig & Laman, 2004; Moore et al., 2013). Reversal of cholesterol loading can restore the migration ability of macrophages (Qin et al., 2006). Calponin 2 is a regulator of cell motility via the binding to F-actin and inhibition of actin-activated MgATPase activity of myosin II (Abe et al., 1990; Winder & Walsh, 1990). This function plays a role in modulating smooth muscle contractility and corresponds to the effect of calponin 2 on stabilizing the actin cytoskeleton in non-muscle cells and inhibiting cell motility (R. Liu & Jin, 2016a). Calponin 2 and myosin II are both concentrated in the center of the cell body of macrophages, supporting the hypothesis that deletion of calponin 2 removes an inhibition of myosin II motor to increase the dynamics of the cytoskeleton. This mechanism lays the foundation for calponin 2 to regulate actin cytoskeleton-based functions, such as cell proliferation, adhesion and migration (Hossain et al., 2003; Huang et al., 2008; R. Liu & Jin, 2016b; Moazzem Hossain et al., 2014; Qian et al., 2021).

Studies have demonstrated that primary fibroblasts and peritoneal macrophages isolated from *CNN2*$^{-/-}$ mice migrated faster than that of WT control cells (Hossain et al., 2005; Huang et al., 2008). The intrinsic motility of calponin 2-null macrophages and foam cells were then investigated in the absence of chemotactic stimulation. In vitro wound healing assay in monolayer cell cultures measured the rate of two-dimensional migration of thioglycolate-elicited mouse peritoneal macrophages in which the level of calponin 2 expression is similar to that in peritoneal residential macrophages as shown by Western blotting. The results showed a significantly faster closure of the scratch wound in *CNN2*$^{-/-}$ macrophages versus WT control, consistent with the faster than WT migration velocity of *CNN2*$^{-/-}$ macrophages measured on individual cells by tracking cell movement using time-lapse microscopy (Huang et al., 2016). The migration velocities of WT and *CNN2*$^{-/-}$ foam cells were both significantly hindered as compared to that of the genotype-matched macrophages. However, the migration velocity of *CNN2*$^{-/-}$ foam cells was hindered significantly less than that of WT foam cells. As a consequence, *CNN2*$^{-/-}$ foam cells moved even faster than that of WT macrophages without lipid ingestion (R. Liu & Jin, 2016b).

Since *CNN2*$^{-/-}$ and WT macrophages show similar degrees of lipid loading, the less impaired motility of *CNN2*$^{-/-}$ foam cells is likely based on higher intrinsic cytoskeleton dynamics other than increased lipid efflux. This notion is supported by the fact that the motility of *CNN2*$^{-/-}$ foam cells remained more active even when compared to that of WT macrophages without lipid loading (R. Liu & Jin, 2016b). The finding that deletion of calponin 2 increases the motility of not only macrophages but also foam cells by overcoming the negative impact of lipid loading indicates that calponin 2 is a potential target for enhancing the motility of macrophages and foam cells to attenuate the progression of atherosclerosis. In contrast to therapeutic manipulations of cellular signaling pathways, deleting calponin 2 to increase the motility of macrophages and compensate for the hindered motility of foam cells presents a downstream effector level mechanism to provide a targeted treatment for atherosclerosis with less side effects as shown by the effective attenuation of hyperlipidemia-induced atherosclerosis in *ApoE*$^{-/-}$,*CNN2*$^{-/-}$ mice.

10.4. Deletion of Calponin 2 in Macrophages Attenuates the Development of Atherosclerosis with Reduced Production of Inflammatory Cytokines

Cytokine-mediated cell signaling plays dominant roles during the pathogenesis and progression of atherosclerosis (Ramji & Davies, 2015). In comparison with untreated WT macrophages, WT foam cells showed increases in the production of cytokines associated with monocytosis (M-CSF and IL-3) and inflammation (IL-1β and VEGF). On the other hand, untreated *CNN2*$^{-/-}$ macrophages had decreased baseline levels of cytokines associated with monocytosis (G-CSF and M-CSF) and inflammation (IL-6, IFN-γ and CXCL10). In comparison with that of WT foam cells, *CNN2*$^{-/-}$ foam cells produced significantly lower levels of cytokines associated with monocytosis (Eotaxin, G-CSF, M-CSF and IL-3) and inflammation (IL-6, IL-12, IFN-γ, CXCL10, CXCL1, TNF-α and VEGF) (R. Liu & Jin, 2016b).

IL-1β is a prominent pro-inflammatory cytokine produced by macrophages following ingestion of oxidized LDL (Freigang et al., 2013; Kamari et al., 2011). Atherosclerotic lesions in *ApoE*$^{-/-}$ mice transplanted with *IL-1β*$^{-/-}$ bone marrow cells were reduced to half in comparison to that in *IL-1β*$^{+/+}$ bone marrow transplanted controls (Kamari et al., 2011). Atherosclerosis development is also accompanied by the up-regulation of VEGF (Kimura et al., 2007) through stimulating the proliferation and growth of endothelial cells, inducing angiogenesis, and potentially promoting plaque formation and destabilization (Holm et al., 2009). M-CSF and IL-3, cytokines that are associated with monocytosis, are induced by hypercholesterolemia and cause proliferation of hematopoietic stem cells and progenitor cells (Yvan-Charvet et al., 2010). Comparing to WT controls, deletion of calponin 2 decreases the productions of monocytosis associated cytokines G-CSF, M-CSF and IL-3 and pro-inflammatory cytokines IL-6, and IFN-γ in foam cells as well as in macrophages, indicating a baseline anti-inflammatory phenotype that effectively overrides the pro-inflammatory stimulation of lipid ingestion. This mechanism provides a molecular basis for the attenuated development of atherosclerosis in *Apo*E$^{-/-}$,*Cnn2*$^{-/-}$ and *Apo*E$^{-/-}$,*Cnn2*$^{f/f}$,*lysM*$^{cre+}$ mice (R. Liu & Jin, 2016b).

10.5. Deletion of Calponin 2 Decreases Cell Adhesion as a Potential Mechanism to Reduce Pro-Inflammatory Activation of Macrophages

It has been broadly observed that substrate adhesion is critical for macrophage differentiation (H. Liu et al., 2008; Szekanecz & Koch, 2007). Macrophages cultured on stiffer substrate exhibited increased spreading area and enhanced adhesion, accompanied with elevated classical activation than that of macrophages cultured on softer substrate (Blakney et al., 2012). Macrophage grown on soft substrates produced less proinflammatory cytokines with decreased TLR4 activity than that of the macrophages grown on rigid substrates (Previtera & Sengupta, 2015). The modulation of macrophage function by substrate rigidity is dependent on actin polymerization and Rho GTPase activation (Patel et al., 2012). Calponin 2 plays a role in enhancing the adhesion of cells to culture substrate (Moazzem Hossain et al., 2014). Similar to the findings in other cell types (Hines et al., 2014; Hossain et al., 2005, 2006; Moazzem Hossain et al., 2014), calponin 2 facilitates the adhesion of macrophages to the culture substrate. Deletion of calponin 2 significantly slowed and weakened adhesion of macrophages to the culture substrate, suggesting a mechanism for calponin 2 to regulate macrophage functions via altering cell adhesion (R. Liu & Jin, 2016b).

To investigate the mechanism for calponin to regulate cell adhesion, primary skin fibroblasts were studied taking advantage of their extended spreading in culture, which permits more clear imaging of the cytoskeleton than in macrophage cultures (R. Liu & Jin, 2016b). Typical actin stress fibers were seen in cultured neonatal mouse skin fibroblasts. Immunofluorescence staining using an anti-calponin 2 mAb 1D11 and a rabbit polyclonal antibody RAH2 revealed co-localizations of calponin 2 with F-actin, tropomyosin and myosin IIA, but not with paxillin-stained focal adhesions (Turner, 2000). Calponin 2 is concentrated in the center of the cell body as that of myosin II. These results suggest that calponin 2 enhances substrate adhesion of cells by decreasing dynamics of the actin cytoskeleton through the inhibition of myosin motor function, a fundamental function of calponin (Haeberle, 1994; Shirinsky et al., 1992). The deletion of calponin 2 to increase the dynamics of actin cytoskeleton and weaken cell adhesion may be responsible for the decreased baseline production of pro-inflammatory cytokines in *CNN2*$^{-/-}$ macrophages and the attenuated up-regulation of pro-inflammatory cytokines in *CNN2*$^{-/-}$ foam cells (R. Liu & Jin, 2016b).

Most of the available knowledge of macrophage differentiation and functions in inflammatory diseases is mainly from studies of receptor-ligation based signaling pathways. The findings that calponin 2, a cytoskeleton regulatory protein, effectively modifies the function of macrophages in the development of atherosclerosis (R. Liu & Jin, 2016b) and rheumatoid arthritis (Huang et al., 2016), demonstrate a cell motility-based novel mechanism to attenuate inflammatory diseases. The data provide evidence that changes in myosin motor-based cytoskeleton dynamics and cell adhesion can alter macrophage activities, suggesting a promising therapeutic target for the treatment and prevention of various inflammatory diseases.

Chapter 11

Repressed Expression of Calponin 2 is Critical to Physiological Quiescence of Lung Macrophages

Aberrant activation and polarization of macrophages are associated with a wide range of inflammatory and autoimmune diseases (Y. C. Liu et al., 2014). Targeting macrophage functions is a promising direction for therapeutic interventions. Macrophage polarization towards pro-inflammatory (e.g., M1) or pro-resolution (e.g., M2) phenotype is directed by both soluble factors such as cytokines and growth factors and mechanical cues such as strain and stiffness of the extracellular matrix (McWhorter et al., 2015; Sridharan et al., 2015). Macrophage activation also involves changes in cell morphology and cytoskeletal rearrangements leading to phenotype differentiations (Martin et al., 1995). Actin-dependent cell elasticity was recognized to determine phagocytosis and cytokine secretion by macrophages (Patel et al., 2012). To accentuate the role of cytoskeleton in macrophage function, directly altering macrophage cell shape was shown to initiate macrophage polarization in an actomyosin contractility-dependent manner (McWhorter et al., 2013). Therefore, manipulating actin cytoskeleton, e.g., via calponin regulations, to induce functional changes in macrophages may lead to innovative therapeutic approaches against the pathogenesis and progression of inflammatory diseases.

Resident macrophages in different organs and tissues exhibit different immunophenotypes and function adept to particular tissue microenvironments (L. C. Davies et al., 2013; Gordon & Pluddemann, 2017). Lung macrophages are uniquely located at a boundary with the outside environment where they are constantly exposed to new antigens and must sustain tolerogenicity while maintaining capacity to initiate inflammation in response to infection (Kopf et al., 2015; Lambrecht, 2006). In contrast, peritoneal cavity is normally sterile with no direct interaction with the external environment, exposure of peritoneal macrophages to inflammatory stimuli provokes a rapid response (Fieren, 2012; J. Wang & Kubes, 2016). To determine whether calponin 2 contributes to the tissue-specific adaptation and activation of macrophages in a mechanically dynamic tissue microenvironment (Roan & Waters, 2011; Tschumperlin et al., 2010), a study examined the role of calponin 2 in lung alveolar resident macrophages (referred to as lung macrophages hereafter) (Plazyo et al., 2019).

11.1. Lung Macrophages Express Significantly Lower Levels of Calponin 2 than That in Peritoneal Macrophages

The levels of calponin 2 in mouse lung macrophages were compared with that in peritoneal macrophages. Since expression of calponin 2 may vary with changes in actin cytoskeleton (Hossain et al., 2005), densitometry analysis of anti-calponin 2 Western blots was normalized to the levels of histones that reflect the number of cells in addition to that of actin and total protein stained in parallel SDS-gels. The quantifications detected a significantly lower level of calponin 2 in lung macrophages. Flow cytometric analysis further confirmed the significantly lower expression of calponin 2 in $CD45^{+}F4/80^{+}$ lung macrophages with their fluorescent spectra entirely overlapping with that of a negative control (Plazyo et al., 2019).

Freshly isolated lung macrophages are noticeably larger in size than peritoneal macrophages, as depicted by their rightward shift in flow cytometric forward scatter plot. Because larger cells possess more actin cytoskeletal networks, the down-regulation of calponin 2 in lung macrophages normalized to cell number implies that the calponin:actin ratio is further reduced in lung macrophages, indicating a mechanism that is known to affect the mechanics of actin filaments in vitro (Jensen et al., 2014). The data that the level of calponin 2 in lung macrophages is markedly lower than that in peritoneal macrophages indicate potentially significant functional effects (Plazyo et al., 2019).

11.2. Lower Level of Calponin 2 Leads to Reduced Adhesion and Correlates with Decreased Expression of F4/80

Calponin binds the key components of cytoskeleton, F-actin and myosin (EL-Mezgueldi & Marston, 1996), and regulates cell adhesion by inhibiting myosin motor-driven cytoskeletal dynamics and motility (J.-P. Jin et al., 2008; R. Liu & Jin, 2016a; K. G. Morgan & Gangopadhyay, 2001). Reduction in calponin

2 is known to result in decreased adhesion of prostate cancer cells (Moazzem Hossain et al., 2014), platelets (Hines et al., 2014), fibroblasts (Hossain et al., 2016), and peritoneal macrophages (Huang et al., 2008; R. Liu & Jin, 2016b). Similarly, lung macrophages expressing lower levels of calponin 2 adhere less efficiently to culture substrate than that of peritoneal macrophages. Calponin 2-null peritoneal macrophages isolated from *CNN2* KO mice also show reduced adhesion like that of WT lung macrophages. No significant difference in substrate adhesion was found between WT and *CNN2* KO lung macrophages which have very low or null calponin 2 contents. Therefore, a threshold of calponin 2 may be required to enhance cell adhesion and lung macrophages restrict the expression of calponin 2 to reduce cell adhesion (Plazyo et al., 2019).

F4/80 is a cell surface glycoprotein extensively used as a marker of murine macrophages. The biological role of F4/80 in macrophages remains not well understood, but may involve cell-cell and cell-matrix contacts (Gordon et al., 2011). The expression of F4/80 is modulated by the state of macrophage activation (Haidl & Jefferies, 1996; Leal et al., 2013; Yu et al., 2016) and increases upon culturing on and adherence to plastic culture dishes in vitro (Hirsch et al., 1981), which are cellular processes that rely on the functions of actin cytoskeleton (Schwarz & Gardel, 2012) and involve calponin 2 regulations (Hossain et al., 2005). It was interesting that F4/80^{hi} peritoneal macrophages are calponin 2-positive while F4/80^{lo} peritoneal macrophages are calponin 2-negative (Plazyo et al., 2019). F4/80^{lo} peritoneal macrophages are derived from monocytes constitutively entering the peritoneal cavity and serve as precursors to F4/80^{hi} residential macrophages (Bain et al., 2016). Therefore, the phenotypic adaptation of newly arrived peritoneal macrophages to the peritoneal environment may involve an up-regulation of calponin 2. Consistent with the positive correlation between the expressions of calponin 2 and F4/80, *CNN2* KO peritoneal macrophages exhibit decreased expression of F4/80 (Plazyo et al., 2019).

Lung macrophages of WT mice are both F4/80^{lo} (L. C. Davies et al., 2013) and *CNN2*^{lo} (Plazyo et al., 2019), a phenotype which corresponds to the observation that lung macrophages display weaker adhesion in response to pro-inflammatory stimuli when compared to other myeloid cells, including blood monocytes (Lundahl et al., 1996). A study assessing the production of superoxide showed that the adhesion of lung macrophages directly correlates with their activation (Ryer-Powder & Forman, 1989). Therefore, lung macrophages may employ the reduced expression of calponin 2 to decrease cell adhesion and subsequently lower the expression of F4/80 (Plazyo et al., 2019) to restrict their inflammatory response.

11.3. Lower Levels of Calponin 2 Reduce the Abundance of Pro-Inflammatory ($CD38^+$, $iNOS^+$) Macrophages with No Change in Pro-Resolution ($CD206^+$, $Egr2^+$) Macrophages

To determine whether the low expression of calponin 2 in lung macrophages restricts pro-inflammatory polarization, immunophenotyping of freshly isolated macrophages using flow cytometry showed that the expression of pro-inflammatory M1 mediators such as nitric oxide synthase (iNOS) (Rath et al., 2014) and cyclic ADP ribose hydrolase (CD38) (Jablonski et al., 2015) demonstrated that the percentage of $CD38^+iNOS^+$ macrophages was significantly lower in lung macrophages than that in peritoneal macrophages. Peritoneal macrophages isolated from *CNN2* KO mice also contained fewer $CD38^+iNOS^+$ macrophages than that from WT mice. No difference in the expression of CD38 and iNOS was found between WT and *CNN2* KO lung macrophages, suggesting that the low levels of calponin 2 in WT lung macrophages are already below the threshold for facilitating macrophages to polarize towards M1-like phenotype (Plazyo et al., 2019).

To further assess whether calponin 2 affects macrophage polarization towards alternative pro-resolution phenotype, expression of M2 markers, mannose receptor (CD206) (Roszer, 2015) and early growth response protein 2 (Egr2) (Jablonski et al., 2015), were evaluated in live lung and peritoneal macrophages isolated from WT and *CNN2* KO mice. While the percentages of $CD206^+$ and $Egr2^+$ macrophages were higher among lung macrophages than peritoneal macrophages, the deletion of calponin 2 did not have an effect, indicating that the upregulation of M2 markers in lung macrophages is independent of calponin 2 or the low calponin 2 in WT lung macrophages has already maximized the upregulation. Taken together, the immunophenotyping data demonstrate that the low expression of calponin 2 abates pro-inflammatory M1-like phenotype and may steer M2-like phenotype in lung macrophages (Plazyo et al., 2019), confirming that calponin 2 promotes pro-inflammatory activation of macrophages (Huang et al., 2016).

11.4. Deletion of Calponin 2 Results in Decreased Expression of IL-1α, IL-1β, IL-6 and TNF in Peritoneal Macrophages and Attenuates Pro-Inflammatory Polarization

The lower abundance of pro-inflammatory peritoneal macrophages in *CNN2* KO mice than that in WT mice (R. Liu & Jin, 2016b) suggests that low or deletion of calponin 2 modulates macrophage activation and promotes quiescence. Quantification of the intracellular protein concentrations of several cytokines using bead-based multiplex immunoassays to compare the activation of peritoneal macrophages in *CNN2*$^{-/-}$ and WT mice showed that consistent with their abated M1 phenotype, *CNN2*$^{-/-}$ peritoneal macrophages produced less pro-inflammatory cytokines IL-1α, IL-1β, IL-6 and TNF than that of WT macrophages (Plazyo et al., 2019). Peritoneal macrophages are known to synthesize basal levels of pro-inflammatory cytokines (Rana et al., 1996) that can be quickly upregulated and secreted in response to pathogen-associated molecular patterns or endogenous damage-associated molecular patterns (H. Kumar et al., 2011; X. Zhang & Mosser, 2008). The results demonstrate that the deletion of calponin 2 decreases the basal levels of pro-inflammatory cytokines in peritoneal macrophages, potentially constraining inflammatory responses (Plazyo et al., 2019). These findings are consistent with the reports that calponin 2-null peritoneal macrophages and lipid-loaded foam cells have reduced expression of inflammatory cytokines and chemokines (R. Liu & Jin, 2016b) and that myeloid cell-specific deletion of calponin 2 attenuates the severity of inflammatory arthritis (Huang et al., 2016) and atherosclerosis (R. Liu & Jin, 2016b) in mice.

11.5. Reduction of Calponin 2 Attenuates Inflammatory Response of Macrophages to Silica Stimulation

Crystalline silica is considered a damage-associated molecular pattern (Busso & So, 2012; G. Y. Chen & Nunez, 2010) that gets internalized by macrophages (Gozal et al., 2002) and activates nuclear factor-κB (Di Giuseppe et al., 2009; J. L. Kang et al., 2000) and inflammasome (Dostert et al., 2008; Franchi et al., 2009; Hornung et al., 2008) signaling. To determine whether calponin 2 contributes to the inflammatory response of macrophages upon treatment with pro-inflammatory stimuli, mouse lung and peritoneal macrophages were cultured for 72 hours with or without 100 μg/mL 200-nm silica particles. Silica treatment resulted in increases of IL-10 and TNF proteins in both lung and peritoneal macrophages isolated from WT or *CNN2*$^{-/-}$ mice. However, the silica-induced increase in IL-10 protein expression was significantly higher in WT peritoneal macrophages than WT lung macrophages or *CNN2*$^{-/-}$ peritoneal macrophages (Plazyo et al., 2019). A similar trend was also evident for TNF, although the quantification did not reach the threshold of statistical significance. No differences in silica-induced expression of IL-10 or TNF were apparent between WT and *CNN2*$^{-/-}$ lung macrophages, as predicted based on their aforementioned functional and immunophenotypic similarities. Concurrent incubations with fluorescently labeled silica confirmed that the internalization of silica by all of the four macrophage groups reached saturation during the 72-hour incubation and, therefore, the effect of calponin 2 on the capacity of phagocytosis (Huang et al., 2008) is deemed to be a negligible factor in the observed response (Plazyo et al., 2019).

11.6. Calponin 2 Participates in Endoplasmic Reticulum Stress Response

Pro-inflammatory polarization of macrophages involves activation of endoplasmic reticulum (ER) stress signaling and autophagy (Bronner et al., 2015; P. Chen et al., 2014; Minton, 2017). Actin cytoskeleton dynamics are linked to ER stress responses (Chambers et al., 2015; Ishiwata-Kimata et al., 2013; St-Pierre et al., 2017; van Vliet et al., 2017). The cross-talk between ER stress and inflammation is also well-documented (Dandekar et al., 2015; Grootjans et al., 2016; Reverendo et al., 2019). As an actin cytoskeleton-associated regulatory protein, calponin 2 may contribute to immune regulation by participating in ER stress signaling. Indeed, flow cytometry (Popat et al., 2018; Warnes, 2014) study showed that *CNN2*$^{-/-}$ peritoneal macrophages exhibit significantly reduced expression of ER stress markers PERK and CHOP than that in WT peritoneal macrophages (Plazyo et al., 2019). Consistently, WT lung macrophages that are low in calponin 2 showed down-regulated PERK and CHOP. Immunoblotting further showed that the expression of another ER stress marker, GRP78, was lower in WT lung macrophages, *Cnn2*$^{-/-}$ peritoneal macrophages and *CNN2*$^{-/-}$ lung macrophages when compared to that in WT peritoneal macrophages. In contrast, IRE1α was lower in lung macrophages than peritoneal macrophages isolated from both

WT and *CNN2*$^{-/-}$ mice, indicating a calponin 2-independent intrinsic difference of the two populations (Plazyo et al., 2019).

ER stress can provoke recycling of macromolecules and organelles, a process known as autophagy (Hoyer-Hansen & Jaattela, 2007; Senft & Ronai, 2015; Song et al., 2018). However, flow cytometry showed that macrophages deficient in calponin 2 (*CNN2*$^{-/-}$ macrophages and WT lung macrophages) did not significantly differ from the high calponin 2 WT peritoneal macrophages in their expression of a classic autophagy marker LC3B. Immunofluorescence microscopy also revealed no notable difference in intracellular distribution patterns of LC3B among lung and peritoneal macrophages of *CNN2*$^{-/-}$ and WT mice, although p62, a target of autophagy that forms speckles in metabolically stressed cells (Moscat & Diaz-Meco, 2009), exhibited fewer punctate in *CNN2*$^{-/-}$ macrophages than that in WT macrophages (Plazyo et al., 2019). These data suggest that perturbed ER stress singling may be one of the molecular mechanisms by which down-regulation of calponin 2 promotes macrophage quiescence.

11.7. Calponin 2 Is a Novel Regulator of Macrophage Activation and Polarization

Actin cytoskeleton is a key mediator of macrophage activation and polarization (Eswarappa et al., 2008; McWhorter et al., 2013; Pergola et al., 2017). While macrophage adhesion (Chong et al., 1987), migration (Jones, 2000) and phagocytosis (May & Machesky, 2001) clearly rely on the function of cytoskeleton, other mechanisms by which cytoskeletal dynamics affect macrophage activity have also been investigated (Mostowy & Shenoy, 2015; Stow & Condon, 2016). Rapid cytoskeletal rearrangements allow for reorganization and clustering of pattern recognition receptors on actin filaments-supported membrane protrusions that act as platforms for juxtaposing receptors and intracellular signaling machinery at the sites of contact with pathogens (Inoue & Shinohara, 2014; Stow & Condon, 2016; St-Pierre et al., 2017). Actin cytoskeleton physically interacts with and affects the activity of inflammasomes and intracellular oligomers that process the pro-inflammatory cytokines IL-1β and IL-18 into their mature forms (Burger et al., 2016; M. L. Kim et al., 2015; Man et al., 2014; Mostowy & Shenoy, 2015; Waite et al., 2009).

Providing the first line of defense against microbial invasion while receiving constant exposure to extrinsic antigens, lung macrophages need to maintain a necessary level of activity while limiting exaggerated inflammatory reaction. Therefore, their low level of calponin 2 may reflect an important physiological adaptation. The different levels of calponin 2 in different macrophage populations may have resulted from their adaptation to specific tissue environment corresponding to specific functional states.

The expression of *CNN2* gene and the degradation of calponin 2 protein are both mechanical tension dependent (Hossain et al., 2005, 2006; Jiang et al., 2014). Lower average mechanical tension that the cellular structures sense results in lower level of *CNN2* gene expression and also promotes calponin 2 degradation. The lung is an organ that continuously experiences a mechanical environment of cyclic stretching and relaxation. Cells adapted to chronic cyclic stretching, such as those in the lung, have had structural remodeling in reference to the longest stretched dimensions, thus sense lower tension during the relaxation phases (Hossain et al., 2006; Rasmussen & Jin, 2024). This mechanism may result in a low average cytoskeleton tension over time and be responsible for maintaining the low calponin 2 level in lung alveolar residential macrophages (see Chapter 16 of this book for more details) for their physiological quiescence.

Considering that calponin 2 is an actin cytoskeleton-associated regulatory protein whose removal steers the function and polarization of macrophages to attenuate inflammatory arthritis and atherosclerosis (Huang et al., 2106; R. Liu & Jin, 2016b), altering cytoskeletal dynamics by down-regulation of calponin 2 to modulate the processing and secretion of cytokines and promote quiescence could be targeted as a novel approach to treating inflammatory diseases. Based on the roles of calponin 2 in cell migration (Huang et al., 2008; Moazzem Hossain et al., 2014; J. Tang et al., 2006; Ulmer et al., 2013), adhesion (Hines et al., 2014; Hossain et al., 2016; Huang et al., 2008; R. Liu & Jin, 2016b; Moazzem Hossain et al., 2014) and phagocytosis (Das et al., 2014; Huang et al., 2008), further studies on whether calponin 2 regulates pattern recognition receptors clustering, inflammasome activation or other elements of cytoskeletal dynamics in macrophages and how it can be targeted pharmaceutically to modulate inflammation may ultimately establish calponin 2 as a novel molecular target for specifically restricting M1 polarization of macrophages for the treatment of inflammatory diseases.

Chapter 12

Systemic Deletion of Calponin 2 Alleviates Aortic Valve Calcification

Calcific aortic valve disease (CAVD) is the most common heart valve disease in Western societies (Mozaffarian et al., 2016) affecting approximately 13% of the US population (Nasir et al., 2008) and 25% of people 65 or older (Iung & Vahanian, 2011; O'Brien, 2006). It initiates with fibrotic thickening and progresses to extensive calcification of the aortic valve cusps, causing aortic sclerosis and stenosis (Otto, 2008) and subsequent heart failure (Lerman et al., 2015). Aortic valve replacement is currently the only treatment available (Iung et al., 2003; Nishimura et al., 2017). However, heart failure continues to progress after valve replacement, reducing the benefit of surgical treatment (Aziminia et al., 2023). The development of non-surgical therapeutics is hindered by the lack in clear understanding of the pathogenic mechanisms (Dweck et al., 2012; Yutzey et al., 2014).

CAVD does not simply result from a degenerative process (Rajamannan et al., 2011) but involves active cellular mechanisms that could be potentially targeted for treatment (Hutcheson et al., 2014; Miller et al., 2011; Yutzey et al., 2014). Aortic valve interstitial cells (AVICs) are the major cell population in the aortic valve whose pathological activation and differentiation into myofibroblasts and osteogenic cells lead to calcification (Rutkovskiy et al., 2017). The contractile myofibroblasts reorganize the extracellular matrix, promoting apoptosis of AVICs and nucleation of calcium crystals around the apoptotic cell aggregates, matrix vesicles, collagen fibrils and other solid particles (Farzaneh-Far et al., 2001). Osteoblast-dependent calcification in the aortic valve mirrors physiological osteogenesis orchestrated by Runt-related transcription factor 2 (Runx2) and bone morphogenetic proteins (BMPs) (Bostrom et al., 2011; Caira et al., 2006; Towler et al., 2006; Wirrig & Yutzey, 2014; X. Yang et al., 2009) and involves deposition of calcium crystals in structures similar to lamellar bones (Bertazzo & Gentleman, 2017; Mohler et al., 2001).

CAVD and atherosclerosis share many pathogenic processes (Aikawa et al., 2007) with common risk factors such as hypercholesterolemia, smoking, hypertension, diabetes mellitus, and chronic renal disease (Ix et al., 2007; Sathyamurthy & Alex, 2015; Stewart et al., 1997). With its role in regulating cytoskeleton functions in macrophages and fibroblasts (R. Liu & Jin, 2016a), calponin 2 contributes to the cellular mechanoregulation in the pathogenesis of CAVD (Plazyo et al., 2018).

12.1. Systemic, But Not Myeloid-Specific, Deletion of Calponin 2 Highly Effectively Attenuates the Development of CAVD in ApoE KO Mice

Suggesting a role of calponin 2 in the pathogenesis of CAVD, human calcified aortic valves exhibit higher expression of calponin than that in healthy valves (Latif et al., 2015), and activation of AVICs coincides with increased expression of calponin (Porras et al., 2017).

*Apo*E plays a key role in cholesterol transportation and metabolism (Dose et al., 2016). *Apo*E KO mouse is an established animal model for atherosclerosis and for CAVD, as dysregulated lipid metabolism is a risk factor for both conditions (Sider et al., 2011). Immunofluorescence microscopy found higher expression of calponin 2 in in the aortic valve leaflets from *ApoE* KO mice than that in WT mouse valve leaflets (Plazyo et al., 2018). The aortic root tissue of *Apo*E KO mice also exhibited increased expression of smooth muscle actin (SMA), a standard marker of myofibroblast differentiation (Baum & Duffy, 2011). These observations suggest that calcification of the aortic valve is associated with increased expression of calponin 2 and myofibroblast activation that is a central mechanism in the pathogenesis of CAVD (J. H. Chen et al., 2011; G. A. Walker et al., 2004). Comparing the aortic valve calcification in 6.5 months old *Apo*E KO and *Apo*E,*CNN2* double KO mice by von Kossa histological stain revealed that *CNN2*,*Apo*E double KO mice developed significantly less calcification in the leaflets of the aortic valve than that in *Apo*E single KO group (Plazyo et al., 2018).

In parallel to atherosclerosis (Rattazzi & Pauletto, 2015), inflammation and macrophage functions also play a role in the pathogenesis of CAVD (S. H. Lee & Choi, 2016; X. F. Li et al., 2016; Murashita & Badhwar, 2017). Pathological inflammation involving infiltration of monocytes and their differentiation into macrophages in response to endothelial damage and lipid deposition is a shared mechanism in CAVD and atherosclerosis (Dweck et al.,

2012; Otto et al., 1994). However, in contrast to the significant attenuation of aortic valve calcification by systemic *CNN2* KO, myeloid cell-specific *CNN2* KO did not reduce the aortic valve calcification caused by *Apo*E KO (Plazyo et al., 2018) although it highly effectively attenuated *Apo*E KO-produced arterial atherosclerosis (R. Liu & Jin, 2016b) that shares many pathogenic processes with CAVD. This important finding indicates the role of calponin 2 in non-myeloid cell types in the pathogenesis of CAVD, also consistent with the notion that the pathogeneses of atherosclerosis and CAVD are significantly different (Dweck et al., 2013).

12.2. Expression of Calponin 2 Increases during AVIC Differentiation and Calcification

AVIC is the major cell type in aortic valve, present in all three of its layers (lamina fibrosa, lamina spongiosa and lamina ventricularis) (Rutkovskiy et al., 2017). In healthy aortic valve tissues, AVICs exhibit a quiescent phenotype which switches to an activated phenotype upon tissue injury or disease (A. C. Liu et al., 2007; Rabkin-Aikawa, Aikawa, et al., 2004; Rabkin-Aikawa, Farber, et al., 2004). When AVICs are isolated and cultured on plastic culture dishes that is a high stiffness substrate, they become activated (Porras et al., 2017). In ovine primary AVICs activated in culture on high stiffness plastic dishes for 3 passages (Porras et al., 2017), Western blot analysis and densitometry quantification demonstrated the increasing expressions of calponin 2 and SMA in AVICs cultured on plastic substrate (Plazyo et al., 2018). The effect of calponin 2 on the differentiation of AVICs, rather than on the activation of macrophages, could be a novel mechanism in the pathogenesis of CAVD.

TGF-β1 is a critical paracrine factor for myofibroblast differentiation and calcification in the pathogenesis of CAVD (Ansorge et al., 2017). To demonstrate that calponin 2 participates in TGF-β1-initiated myofibroblast differentiation of AVICs, immunofluorescence studies showed that treatment of AVICs with TGF-β1 in culture for 6 days increased the expression of both calponin 2 and SMA with their incorporation into the stress fibers of cytoskeleton together with increased calcification detected by von Kossa and alizarin red stains (Plazyo et al., 2018).

12.3. Calponin 2 Deletion Attenuates Myofibroblast Differentiation

To examine how down-regulation of calponin 2 may affect myofibroblast differentiation, fibroblasts isolated from *CNN2* KO mice were examined in culture under activating conditions. The upregulation of SMA in response to TGF-β1 was markedly lower in *CNN2* KO fibroblasts than that in WT cells (Plazyo et al., 2018). This finding indicates that calponin 2 may function as a rate-determining facilitator in myofibroblast differentiation.

Demonstrating that the fibroblast-to-myofibroblast differentiation serves as a mechanistic model of calponin 2-regulated calcification, von Kossa staining revealed that treatment with TGF-β1 in culture for 6 days resulted in a striking increase of calcification in WT fibroblasts, which was markedly lessened in *CNN2* KO fibroblasts. Quiescent AVICs exhibit a fibroblast-like phenotype and can differentiate into myofibroblast-like cells upon activation, a process that can be mimicked in vitro by culturing on high-stiffness substrates (H. Ma et al., 2017). It was found that even when cultured on high-stiffness dishes without adding TGF-β1, WT fibroblasts tended to exhibit higher calcification than *CNN2* KO fibroblasts (Plazyo et al., 2018). These observations indicate the role of calponin 2 in facilitating calcification of myofibroblasts cultured under activating conditions, including mechanosignaling such as that from high-stiffness culture substrate.

12.4. Calponin 2 Plays a Role in the Early Stage of Myofibroblast Differentiation

Confirming the role of calponin 2 in myofibroblast activation that is a leading step in the development of CAVD, Western blot analysis showed that the expression of calponin 2 and SMA increased during culture and passaging of primary fibroblasts on high stiffness plastic dishes. In WT fibroblasts, the expression of calponin 2 increased significantly earlier than the expression of SMA during the 72 hours culture of the first passage of cells, whereas the expression of both calponin 2 and SMA plateaued at 24-48 hours of culture of the second passage of cells. The expression of SMA in *CNN2* KO fibroblasts was significantly lower than that in WT cells. These expression patterns suggest that calponin 2 participates in myofibroblast differentiation, potentially acting as an early-stage facilitator in the pathogenesis of CAVD.

The expression of calponin 2 increased during myofibroblast-like differentiation of primary sheep aortic valve interstitial cells and during the osteogenic differentiation of mouse myofibroblasts. TGF-β1 upregulates the expression of calponin 2. Deletion of calponin 2 decreases and delays TGF-β1-induced myofibroblast differentiation and calcification (Plazyo et al., 2018). The early stage facilitator role of calponin 2 in the development of CAVD is consistent with the finding that upon TGF-β1 treatment of cells cultured on high stiffness plastic dishes, the increased expression of calponin 2 reaches a plateau before that of SMA (Plazyo et al., 2018).

12.5. Calponin 2 Plays a Role in Osteogenic Differentiation of AVIC

In addition to myofibroblast-like differentiation, osteoblast-like differentiation is also evident in CAVD (Rajamannan et al., 2003) with 13% of patients possessing true bone with osteoblasts and osteoclasts in calcified aortic valves at the time of surgical valve replacement (Rutkovskiy et al., 2017). AVICs may be differentiated into osteoblasts directly (J. H. Chen & Simmons, 2011) or following their intermediary differentiation into myofibroblasts (Hjortnaes et al., 2016). As an early stage regulator of AVIC activation, calponin 2 may contribute to their differentiation into both myofibroblasts and osteoblasts, which would be consistent with the reports implicating induction of calponin 2 in osteoblast differentiation and mineralization (Kitching et al., 2002; Seth et al., 2000). Further research is needed to establish the role of calponin 2 in osteogenic signaling during the development of CAVD.

12.6. Mechanoregulation of Calponin 2 as a Novel Mechanism in the Pathogenesis of CAVD

Atherosclerosis and CAVD differ in their hemodynamic basis (Lerman et al., 2015). While the blood vessels affected by atherosclerosis are exposed to sustained laminar blood flow, the aortic valve is susceptible to pulsatile shear stress on the ventricular side and reciprocating shear stress on the aortic side (Chiu & Chien, 2011; Lerman et al., 2015), which shapes a unique mechanical signaling (Gould et al., 2013). The aortic valve naturally experiences dynamically changing high mechanical tension under hemodynamics-induced mechanical stimuli that include stretch, shear stress and transvalvular pressure (Arjunon et al., 2013; Balachandran et al., 2011).

Mechanoregulation of AVICs and the cellular responses to tension and tissue stiffness play a central role in the pathogenesis and progression of CAVD (Arjunon et al., 2013; Balachandran et al., 2011; Cirka et al., 2015). AVIC differentiation in CAVD depends on the mechanical tissue environment (J. H. Chen & Simmons, 2011; Hjortnaes et al., 2016; Merryman et al., 2006; 2009, 2009; Porras et al., 2017; Yip et al., 2009). The mechanical adaptation of AVICs is constantly undergoing adjustments in response to the changes in intracellular tension imposed by the external forces (P. F. Davies & Guerraty, 2011). Therefore, identifying a target molecule that regulates the mechanical responsiveness of AVICs may lead to the development of new treatment and prevention of CAVD.

Sustained mechanical tension in the cytoskeleton based on the stiffness of tissue environment is required for AVIC activation (P. F. Davies & Guerraty, 2011). High local shear stress is capable of inducing expression of TGF-β1 (Chiu & Chien, 2011), a critical paracrine factor for myofibroblast differentiation and calcification. Mechanical force lowers the threshold for AVICs' sensitivity to TGF-β1, contributing to their activation (J. H. Chen & Simmons, 2011). The *CNN2* gene expression and calponin 2 protein degradation are regulated by the stiffness of extracellular environment, which in turn modifies the tension in the actin cytoskeleton (Hossain et al., 2005, 2006; Jiang et al., 2014). AVIC activation then progressively stiffens the tissue environment, which upregulates SMA to increases the capacity of the myofibroblasts to generate contractile forces (Sandbo & Dulin, 2011) and osteogenesis to form a positive feedback loop that accelerates calcification (Duan et al., 2016). The function of calponin 2 as a mechanoregulatory protein in actin cytoskeleton (Rasmussen & Jin, 2024) showed a dominant effect on the positive feedback in AVIC activation in the development of CAVD (Plazyo et al., 2018).

TGF-β1 upregulates the expression of calponin 2. TGF-β1 is secreted in an inactive complex that binds the components of extracellular matrix (Annes et al., 2003). Stretching of the tissue and increased contractility of myofibroblasts results in release of biologically active TGF-β1 from the latent reservoir, which is a process requiring SMA-positive stress fibers, cytoskeleton contractile force, and high-stiffness substrate (Wipff et al., 2007). Calponin 2 is associated with actin stress fibers (Danninger & Gimona, 2000; Hossain et al., 2003, 2005; R. Liu & Jin, 2016a), contributes to the regulation of vascular smooth muscle contractility (Feng et al., 2019), increases in expression when cells are cultured on high-stiffness substrates (Hossain et al., 2005; Moazzem Hossain et al., 2014), and regulates cytoskeletal tension in substrate stiffness-dependent manner (Rasmussen & Jin, 2024). These functions of calponin 2 support its contribution to myofibroblast differentiation under mechanoregulated TGF-β1 signaling.

The function of calponin 2 in mechanoregulation is a promising mechanism underlying the effect of calponin 2 deletion on attenuating the development of CAVD. Because calponin 2 acts as a direct modulator of actin and actin-dependent cellular functions, targeting calponin 2 may provide a way to manipulate cytoskeleton activity for altering mechanoregulation in the treatment of diseases involving intrinsically dysregulated mechanical signaling, such as CAVD (Plazyo et al., 2018). The increases of calponin 2 in AVICs and fibroblasts during myofibroblast differentiation and calcification in conjunction with the attenuation effects of calponin 2 deletion present calponin 2 as a novel regulator of pathogenic differentiation of AVICs and a promising molecular target for the treatment and prevention of CAVD.

Chapter 13

Systemic Deletion of Calponin 2 Reduces Postoperative Peritoneal Adhesion

Surgical procedures are invasive and produce tissue lesions that promote inflammatory responses followed with repair including fibrotic remodeling. Postoperative peritoneal adhesion is a common clinical complication caused by operational injuries during abdominal surgeries, including sterile procedures (van Goor, 2007). The formation of peritoneal adhesion post-abdominal surgeries is of high occurrence, relating to the extent and frequency of surgical procedures (Tulandi et al., 2009). The pathogenesis starts with inflammatory responses and wound healing, leading to the formation of fibrotic adhesion between abdominal organs (Capobianco et al., 2017; Mutsaers et al., 2016). Such adhesion is nearly always permanent and constitutes a major health problem, subjecting the patients to a lifelong risk of various secondary disorders (Liakakos et al., 2001). The iatrogenic complications include chronic pain, small bowel obstruction, difficulties in reoperation and female infertility (ten Broek et al., 2016). New surgical techniques such as minimally invasive surgery and anti-adhesion barriers have had limited effects on preventing adhesion formation (Strik et al., 2019).

The pathogenesis of peritoneal adhesion after unavoidable traumatic tissue injuries during surgery involves inflammation, tissue repair and remodeling (Arung et al., 2011). During the inflammatory stage, macrophages play a central role in orchestrating innate immune response. Cytokines and growth factors released from the macrophages activate fibroblasts transformed from mesothelial cells, a process known as mesothelial-to-mesenchymal transition (Hinz, 2016) to differentiate into myofibroblasts with collagen secretion and contractile phenotypes. Excessive inflammatory response and fibroblast activation would lead to the formation of permanent fibrosis and adhesions (Koninckx et al., 2016). Therefore, macrophage-mediated inflammatory response and myofibroblast differentiation are major components of the pathogenesis and progression of postoperative peritoneal adhesion. With calponin 2's functions in the development of inflammatory arthritis (Huang et al., 2016), arterial atherosclerosis (R. Liu & Jin, 2016b) and calcific aortic valve disease (Plazyo et al., 2018) via modulating macrophage activation and myofibroblast differentiation, it also plays a role in the pathogenesis of postoperative peritoneal adhesion (T. B. Hsieh et al., 2022).

13.1. Deletion of Calponin 2 in Mice Significantly Reduced Postoperative Peritoneal Adhesion

A reproduceable non-infectious surgical protocol was used to study the postoperative peritoneal adhesion in mice. Avoiding subjectiveness in visual histopathology to determine the severity of adhesion (Kraemer et al., 2014; Whang et al., 2011), an anatomical scoring system was used for an objective quantification of peritoneal adhesions (Y. L. Yang et al., 2018), in which 0 = no adhesion, 1 = minimally notable adhesions causing cecal wall contractures, 2 = adhesions formed in less than 50% of the length of cecum, 3 = adhesions formed in more than 50% of the length of cecum, and 4 = adhesions formed in more than 50% of the length of cecum plus adhesions of cecum to abdominal wall or other organs (T. B. Hsieh et al., 2022). The results showed tight median ± interquartile range (IQR) with few outliers, providing reliable measurements for the severity of postsurgical peritoneal adhesion. Based on the concept that the area of adhesion is proportional to proliferation of inflammatory and fibrotic cells plus collagen deposition, the area to length ratio of adhesions was used as a normalized quantification in histological sections to objectively assess the development of fibrotic remodeling underlying the extents of postoperative peritoneal adhesion to compare WT and *CNN2* KO mice for the effect of calponin 2 deletion on the development of postoperative peritoneal adhesion.

All mice in sham surgery groups did not develop cecal adhesions. The peritoneal adhesion scores of WT and *CNN2* KO mice showed that the formation of adhesion was progressively intensified from postoperative Day 3 to Day 7 in both groups. However, calponin 2 KO mice exhibited significantly lower adhesion score (median = 1.5, IQR = 1.00-2.25) than that of the WT mice (median = 3.5, IQR = 2.75-4.0) on post-surgery Day 3 ($p < 0.02$). The significant difference continued to post-surgery Day 7 with adhesion scores of median = 3.0, IQR = 2.75-3.0 for calponin 2 KO and median = 4.0, IQR = 3.0-4.0 for WT ($p < 0.02$) (T. B. Hsieh et al., 2022).

Normalized quantification of area to length ratio of adhesions in histological sections confirmed the significantly attenuated pathology in

CNN2 KO mice than that in WT mice with the average area to length ratio of the adhesions in *CNN2* KO mice less than half of that of the WT mice, demonstrating the significant effect of calponin 2 deletion on reducing the formation of postoperative peritoneal adhesion (T. B. Hsieh et al., 2022).

13.2. Deletion of Calponin 2 Attenuates Macrophage Recruitment during Postoperative Peritoneal Wound Healing

F4/80 is a specific marker for mouse macrophages. Immunofluorescent staining of F4/80 on frozen sections of adhesion sites 3 days after surgical lesion showed diffuse infiltration of inflammatory cells in the proliferating mesothelial layer of the adhesion sites and in areas with rich blood supply such as cecal mucosa and submucosal layers. Quantification of macrophage density at adhesion sites found a lower percentage of F4/80$^+$ cells in *CNN2* KO samples than that in WT controls (T. B. Hsieh et al., 2022).

Macrophages are the major cell type orchestrating the inflammatory response during the first three days after tissue injury (Koninckx et al., 2016). Activated macrophages recruited to the injury site release cytokines and other inflammatory mediators to activate fibroblast differentiation into myofibroblasts (Namvar et al., 2018). While the inflammatory response in postoperative peritoneal injuries also involve other immune cells such as neutrophils and lymphocytes, studies have shown that myeloid cell-specific deletion of calponin 2 in macrophages effectively attenuates inflammatory arthritis (Huang et al., 2016) and arterial atherosclerosis (R. Liu & Jin, 2016b). Therefore, the function of calponin 2 in macrophage activation and inflammatory response plays an important role in the development of postoperative peritoneal adhesion, and the deletion of calponin 2 in macrophages may attenuate the initial activation of macrophages to reduce the pathogenesis of postoperative peritoneal adhesion.

13.3. Deletion of Calponin 2 Attenuates Myofibroblast Activation during Postoperative Peritoneal Wound Healing

After peritoneal tissue injury, the repairing and regeneration process involves fibrotic remodeling and scar formation (Capobianco et al., 2017). Calponin 2 expresses in fibroblasts (Hossain et al., 2005). The expression of calponin 2 increases during the differentiation of fibroblasts into myofibroblasts, which plays a critical role in the pathogenesis of aortic valve fibrocalcification (Plazyo et al., 2018). Calponin 2 is not expressed in mesothelial cells but will be turned on in myofibroblasts differentiated from mesothelial cells during trauma-induced mesothelial-to-mesenchymal transition.

Calponin 2 is an upstream regulator leading myofibroblast differentiation and deletion of calponin 2 attenuates myofibroblast differentiation and the development and progression of calcific aortic valve disease (Plazyo et al., 2018). Immunofluorescence microscopy on frozen sections of cecal tissue 7 days after surgical lesion detected myofibroblast marker, α-SMA, positive cells accumulated in the deeper layer of adhesion sites. Quantification of the density of α-SMA positive cells showed a lower percentage of α-SMA positive cells in *CNN2* KO samples as compared to WT control (T. B. Hsieh et al., 2022). The study of systemic calponin 2 deletion mice for surgically induced peritoneal injury reproduced the effect on attenuation of myofibroblast differentiation seen in other models to minimize the formation of postoperative adhesions (T. B. Hsieh et al., 2022).

13.4. Medical Importance

Postoperative peritoneal adhesion is a common cause of morbidity, resulting in multiple complications. Macrophage-mediated inflammation and myofibroblast differentiation after tissue injury play central roles in the pathogenesis and progression. Calponin 2 is an actin cytoskeleton regulatory protein in endothelial cells, macrophages and fibroblasts, all of which are key players in the development of postoperative peritoneal adhesion. Deletion of calponin 2 has been shown to attenuate inflammatory arthritis, atherosclerosis and fibrocalcification of aortic valve. The beneficial effect of calponin 2 deletion on attenuating the formation of postoperative peritoneal adhesion in the mouse model showed another potential value as a new therapeutic approach.

Calponin 2 expression is mechanically regulated (Jiang et al., 2014; Rasmussen & Jin, 2024). Mechanical tension and tissue stiffness may contribute to the development and prognosis of postoperative peritoneal adhesion via affecting the expression and function of calponin 2 in various cell types especially myofibroblast differentiation and fibrotic remodeling, which renders calponin 2 an attractive therapeutic target (T. B. Hsieh et al., 2022).

This finding opens a translational direction for further research toward the development of clinical prevention. Reduction of calponin 2 in cells present in and recruited to the injury site by using localized pharmacological approaches may be able to suppress macrophage activation and the downstream activation of myofibroblast differentiation to attenuate peritoneal adhesions.

The experimental evidence further demonstrates that calponin 2 plays a potent role in the process of wound healing in general. The expressions of calponin 2 in epithelial cells, endothelial cells, fibroblasts and macrophages place it at a unique position as a modulator of the process of wound healing in various medical conditions. It merits further investigations on differentially regulate calponin 2 expression and function in each of these cell types and at different stages of healing to achieve the restoration of normal tissue structure with minimized abnormality such as adhesion, scar formation and contracture. Encouragingly, we have shown in a diabetic cornea wound healing study that myeloid cell-specific *CNN2* KO mice had significantly better healing of the injured cornea (our unpublished results). Cell type-targeted deletion or reduction of calponin 2 may be explored as a novel approach to facilitate various cases of wound healing.

Chapter 14

Transgelins

Transgelins are a family of calponin-like cytoskeleton regulatory proteins (T.-B. Hsieh & Jin, 2023). Transgelin was originally found as a 22-kDa protein in chicken gizzard smooth muscle with the name of SM22 (Lees-Miller, Heeley, & Smillie, 1987; Lees-Miller, Heeley, Smillie, et al., 1987; Pearlstone et al., 1987). This isoform of SM22 (SM22α) was rediscovered later under the name transgelin based on its apparent ability to induce gelation of actin filaments in vitro (Shapland et al., 1993). Transgelin/SM22α is one of the most abundant proteins in vertebrate smooth muscles with a high degree of structural similarity to calponin and a function in the stability of actin cytoskeleton like that of calponin (Gimona et al., 2003). Like calponin, three homologous isoforms of transgelin have been identified in vertebrates.

14.1. Evolutionary Lineage of the Three Vertebrate Transgelin Isoforms

Represented by human data, three isoforms of transgelin are present in vertebrates. Transgelin-1 (also named SM22α, p27, WS3-10), transgelin-2 (SM22β) and transgelin-3 (SM22γ, NP25 or NP25) share ~70% sequence identities (J. Liu et al., 2020). The three transgelin isoforms are encoded by three homologous genes. Transgelin-1, SM22α, p27 and WS3-10 are the same protein encoded by *TAGLN* gene (Almendral et al., 1989; Grigoriev et al., 1996; Lawson et al., 1997) that is 5.4-kb in size localized on human chromosome 11 at q23.3 (Camoretti-Mercado et al., 1998). Transgelin-2/SM22β is encoded by *TAGLN2* gene (Grigoriev et al., 1996) located on human chromosome 1q23.2 (Jo et al., 2018). Transgelin-3/NP22/NP25 is encoded by *TAGLN3* gene (Fan et al., 2001; Prinjha et al., 1994) located on human chromosome 3q13.2 (Assinder et al., 2009). While transgelin-1 is primarily a smooth muscle-specific protein, transgelin-3 was first identified and isolated from rat brain and named NP25.

A phylogenetic analysis showed the evolution lineage of vertebrate transgelin isoforms (T.-B. Hsieh & Jin, 2023) in which the divergency pattern depicts that each of the transgelin isoforms is conserved across the fish, amphibian, reptile, avian and mammalian classes whereas the three isoforms have significantly diverged during vertebrate evolution (Figure 6), which is a pattern similar to that in the vertebrate calponin isoforms (see Figure 2 in Chapter 1) (R. Liu & Jin, 2016a) . The phylogenetic data indicate that the *TAGLN2* gene emerged earlier than *TAGLN and TAGLN3* during vertebrate evolution. *TAGLN2* has also diverged more among vertebrate classes while *TAGLN* and *TAGLN3* are relatively conserved (Figure 6).

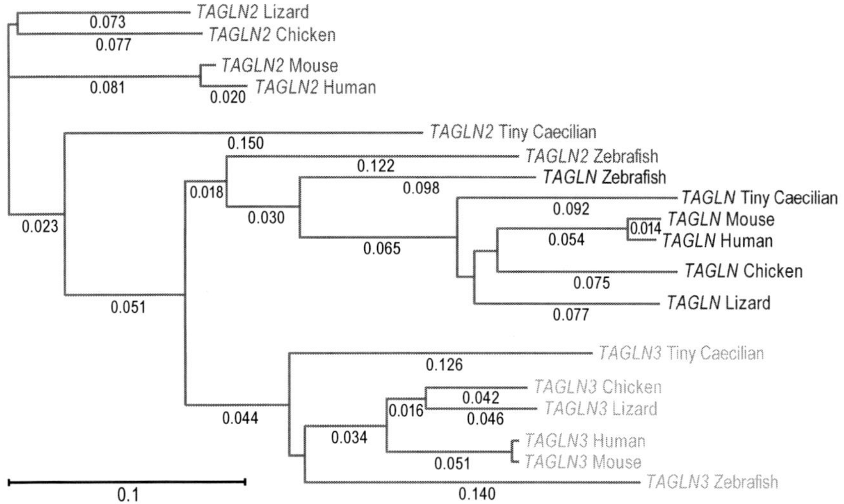

Figure 6. Evolutionary lineage of vertebrate transgelin isoforms. The phylogenetic tree was generated by aligning amino acid sequences of the three transgelin isoforms in representative vertebrate classes including fish, amphibian, reptile, avian and mammal with the MegAlign computer program (Lasergene; DNASTAR, lnc, Madison, WI) using Clustal W method. The degrees of evolutionary divergence are indicated by the lengths of lineage lines. The NCBI database accession numbers of the sequences analyzed are: Zebrafish TAGLN, NP_001038932.1; Zebrafish TAGLN2, AAI65004.1; Zebrafish TAGLN3, AAI51899.1; Tiny caecilian TAGLN, XP_030076731.1; Tiny caecilian TAGLN2, XP_030043052.1; Tiny caecilian TAGLN3, XP_030060064.1; Cape cliff lizard TAGLN, XP_053124973.1; Cape cliff lizard TAGLN2, XP_053133737.1; Cape cliff lizard TAGLN3, XP_053167629.1; Chicken TAGLN, AAA48782.1; Chicken TAGLN2, XP_024999416.1; Chicken TAGLN3, XP_040518015.1; Mouse TAGLN, CAA92941.1; Mouse TAGLN2, NP_848713.1; Mouse TAGLN3, AAH55338.1; Human TAGLN, NP_001001522.1; Human TAGLN2, KAI4083433.1; Human TAGLN3, KAI2530827.1.

14.2. Expression, Function and Regulation of Transgelins

Transgelin-1/SM22α was discovered nearly four decades ago but its biological function remains unclear. Using a specific mAb, a Western blot survey of mouse tissue types showed the expression of transgelin-1/SM22α exclusively in smooth muscles similar to that of calponin 1 (Figure 7) (R. Liu et al., 2017). Therefore, transgelin-1/SM22α has been used as an early differentiation marker of vertebrate smooth muscle cells (Lawson et al., 1997; Zhong et al., 2019). Its levels of expression in smooth muscle cells reflect the degree of differentiation (Kato et al., 2019). Down-regulation of transgelin-1/SM22α is correlated to cell type transformation and tumorigenesis (Dvorakova et al., 2016; Shields et al., 2002). Besides a possible role in regulating smooth muscle contractility, transgelin-1/SM22α has been proposed with functions in regulating the structure and dynamics of actin cytoskeleton and cell motility of fibroblasts (Thompson et al., 2012). The expression pattern of transgelin-1/SM22α in differentiated smooth muscle is similar to that of calponin 1 (Duband et al., 1993) while its regulation of actin cytoskeleton in fibroblasts is similar to that of calponin 2 in non-muscle cells (Assinder et al., 2009; Thompson et al., 2012).

Transgelin-1/SM22α is abundantly expressed in vascular and visceral smooth muscles and early during smooth muscle differentiation (Gimona et al., 2003). However, transgelin-1/SM22α is not required for smooth muscle differentiation although it is involved in calcium-independent smooth muscle contraction (Assinder et al., 2009). It is also expressed in fibroblasts, myofibroblast line MRC-5, and cultured primary skin fibroblasts (R. Liu et al., 2017), and epithelial cells when treated with TGF-β1 to induce myofibroblast differentiation (Elsafadi et al., 2016). Transgelin-1/SM22α was found in some tumor cell lines (Prasad et al., 2010; Schenker & Trueb, 1998) and may act as a tumor suppressor with diminished expression in prostate (Z. Yang et al., 2007), breast (Dvorakova et al., 2016; Sayar et al., 2015), and colon cancers (Assinder et al., 2009; J. Liu et al., 2020). The tumor suppressor function of transgelin-1/SM22α may be related to a suppression of metallomatrix protease-9 that is upregulated in those types of cancer cells (Assinder et al., 2009). With its expression in these non-muscle cell types, plus a mechanical tension-regulated expression (R. Liu et al., 2017), the use of SM22α as universal biomarker for smooth muscle cells and the use of SM22α promoter-driven Cre transgene for inducing smooth muscle-specific gene targeting require a revisit with precaution.

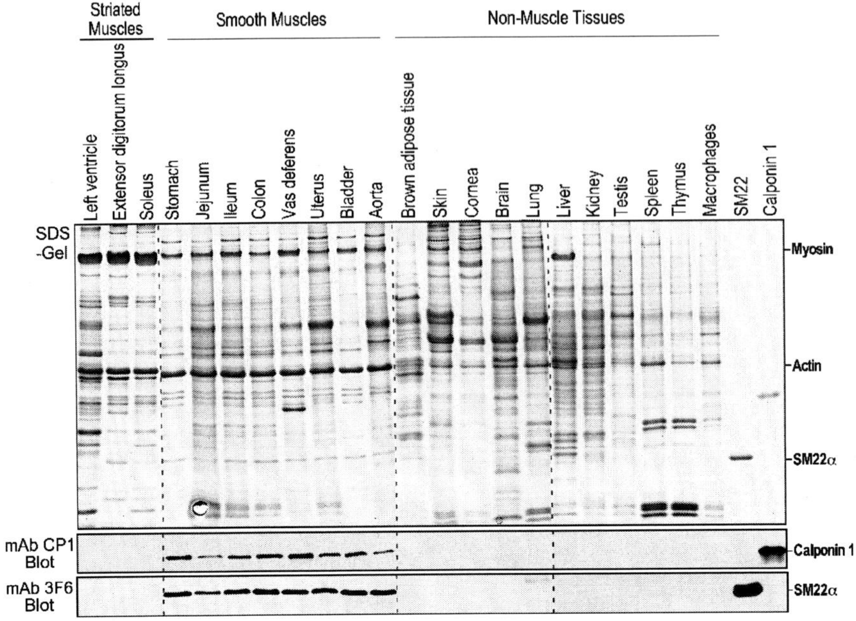

Figure 7. Smooth muscle-specific expression of transgelin-1/SM22α and calponin 1. Total protein extracts from representative organs and tissues of adult mice were analyzed with SDS-PAGE and Western blots using anti-transgelin-1/SM22α mAb 3F6 and anti-calponin 1 mAb CP1. Purified mouse transgelin-1/SM22α and calponin 1 were used as controls. The sample loading was normalized to the level of actin (R. Liu et al., 2017). The result showed the expression of transgelin-1/SM22α and calponin 1 specifically in all smooth muscle organs.

Transgelin-1/SM22α has been found in association with remodeling of the actin cytoskeleton and promotes the migration and invasion of cancer stem cells (J. Liu et al., 2020). Cortical transgelin-2 interacts with light intermediate chain subunit 2-dynein for maintaining proper length of mitotic spindles, chromosome alignment, spindle orientation and timely anaphase onset (A. Sharma et al., 2020). Immunofluorescence staining of monolayer cultures of neonatal mouse primary skin fibroblasts using anti-SM22α, anti-calponin 2 and anti-myosin IIA antibodies together with phalloidin for F-actin demonstrated that transgelin-1/SM22α is co-localized with calponin 2 and the actin-myosin cytoskeleton. The co-localization pattern is consistent with the fact that transgelin-1/SM22α, calponin 2 and myosin IIA are all actin filament-associated proteins with cell motility-related functional interactions (R. Liu et al., 2017).

Transgelin-1 and transgelin-2 have similar functions as actin-binding proteins abundantly expressed in contractile smooth muscle cells (Owens, 1998). They are also expressed in fibroblasts cultured on high stiffness plastic substrates with a role in modulating actin cytoskeleton-based functions such as cell shape, migration and transformation (Dvorakova et al., 2016; Thompson et al., 2012). Like calponin 2, transgelin-2 is expressed in a wide range of cell and tissue types including smooth muscle cells (Yin et al., 2018), epithelial cells, stem cells (Kuo et al., 2011), T lymphocytes (Na et al., 2015), bone marrow cells, and pancreatic tissue (Meng et al., 2017). Transgelin-2 localizes to multiple intracellular compartments, including the cytoplasm, cell membrane and nucleus. The cellular localizations may vary in pathological conditions, such as during epithelial to mesenchymal transition in colorectal cancer (Elsafadi et al., 2020; Lin et al., 2009; Y. Zhang et al., 2010).

14.3. Structure-Function Relationship of Transgelin

Transgelin is phylogenetically homologous to calponin. Transgelin-1 has an isoelectric point of 9.0. Transgelin-2 is also a 22-kDa protein and has an isoelectric point of 8.41 (J. Liu et al., 2020). Transgelin-3 is a protein of 199 amino acids with sequence similarity to other two transgelin isoforms and calponins. Primary structure alignment of transgelin and calponin shows that the entire transgelin structure aligns to the N-terminal and middle regions of calponin with a very high degree of sequence similarity (R. Liu et al., 2017; R. Liu & Jin, 2016a), consistent with the notion that the C-terminal segment of calponin is a variable region that constitutes most of the structural diversity among the three vertebrate calponin isoforms (see Figure 4 in Chapter 2) (R. Liu & Jin, 2016a). The conserved structure of transgelin consists of an N-terminal CH domain, a calcium binding motif with a helix-loop-helix structure, an actin-binding region, and a single CLIK motif (Y. Fu et al., 2000; Kawasaki & Kretsinger, 2017) (Figure 8). The structural and functional similarities demonstrate that transgelins are members of the calponin family proteins and participate in similar functions of actin cytoskeleton. The structural similarity also indicates that transgelin represents the conserved core structure of calponin-transgelin family proteins.

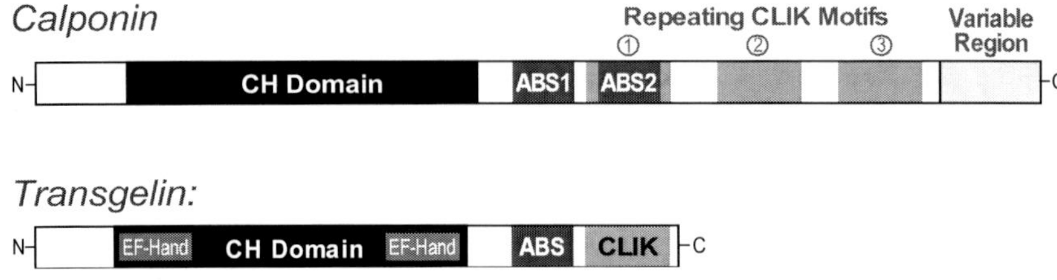

Figure 8. Comparison of the domain architectures of calponin and transgelin. The structural and functional domains of calponin and transgelin are outlined for linear structural comparisons. While calponin has an N-terminal calponin homology (CH) domain, two actin-binding sites (ABS), three CLIK repeats, and a C-terminal variable region that differs among the three isoforms, transgelin contains all the core domain structures of calponin without repeats or the C-terminal variable region. The CH domain of transgelin binds ERK, the ABS domain binds F-actin, and the C-terminal CLIK segment binds ezrin. Two EF-hand Ca^{2+} binding motifs are identified in the CH domain of transgelin, suggesting a possible regulation via calcium signaling.

Studies have indicated that multiple sites in the C-terminal domain of transgelin-1 are required for high affinity actin-binding (Y. Fu et al., 2000). Transgelin-1 possesses EF-hand calcium binding sequences with demonstrated Ca^{2+}-binding activity (Shishibori et al., 1996). Transgelin-3 has conserved sequences in chicken, rat, mouse, and human, containing an N-terminal CH-domain, an actin binding region and a C-terminal CLIK sequence (Ren et al., 1994). In contrast to the three vertebrate calponin isoforms, transgelins lack the C-terminal variable region (T.-B. Hsieh & Jin, 2023). Also shown in Figure 8, transgelin's structural similarity with calponin mainly lies in the CH domain (M. Li et al., 2008) and the first actin-binding site, consistent with their shared feature of actin-binding proteins (Y. Fu et al., 2000). Compared with that of calponin, more regulatory phosphorylation sites are found in transgelin.

14.4. Transcriptional and Posttranslational Regulation of Transgelin-1/SM22α

The gene expression and protein turnover of SM22α are both regulated by mechanical tension. Demonstrating the mechanical tension regulation of SM22α gene expression, the stiffness of culture matrix positively regulates the expression of SM22α mRNA and protein in myofibroblast cell line MRC-5 and primary mouse fibroblasts (R. Liu et al., 2017). This finding in non-muscle cells is supported by smooth muscle studies showing three-day organ culture of rat portal vein in the absence of distension caused lower SM22α synthesis compared with culture under a mechanical load (Zeidan et al., 2000, 2003). Further supporting the regulation by cytoskeleton tension, blebbistatin inhibition of myosin II motor to reduce cytoskeletal tension decreased the expression of SM22α. Blebbistatin treatment needed three days to reduce SM22α, implicating a determining role of transcriptional regulation (R. Liu et al., 2017). The blebbistatin treatment-induced mechanoregulation of SM22α is similar to that of calponin 2, and a transfective expression of calponin 2 using a viral promoter was not under the mechanical tension regulation, demonstrating that transcriptional control is essential (Hossain et al., 2005).

Ex vivo unloading of mechanical tension induced degradation of SM22α in mouse aortic rings, which can be prevented by simply maintaining a tension load. Like that seen with calponin 2 in lung alveola cells (Hossain et al., 2006), this posttranslational regulation by cytoskeleton tension indicates a rapid mechanism to regulate the cellular level and function of SM22α (R. Liu et al., 2017). Mass spectrometry determined that an SM22α fragment produced as the initial intermediate product of low tension-induced degradation is from a specific truncation of the C-terminal 22 amino acids. This cleavage deletes a perspective actin binding site, supporting the notion that this fragmentation is an initial step to free SM22α from actin cytoskeleton to promote the tension-regulated proteolytic degradation. Supporting a Ca^{2+} dependence of SM22α degradation, μ-calpain treatment in vitro reproduced the early cleavage product of SM22α found in tension-unloaded aorta (R. Liu et al., 2017). The C-terminal 22 amino acid segment of SM22α contains a PKC phosphorylation site, possibly homologous and analogous to Ser_{175} in calponin 1 (our unpublished result), suggesting a cell signaling mechanism for further investigations. It is also worth noting that a reported crystal structure of SM22α (M. Li et al., 2008) did not resolve the C-terminal 46 amino acids, implicating that the C-terminal domain is a flexible structure relevant to functions such as actin-binding and posttranslational modifications.

14.5. Transgelin Represents a Prototype of the Calponin Family of Cytoskeleton Regulatory Proteins

The structure of vertebrate transgelins is highly similar to the core structure of calponins (T.-B. Hsieh & Jin, 2023). Transgelin and calponin have conserved primary structures. Antibodies raised by transgelin-1/SM22α immunization have cross-reactions with calponin 1 (R. Liu et al., 2017), indicating conservation of folded structures. The fact that transgelin-1/SM22α aligned with the N-terminal and middle regions that are conserved in all calponin isoforms with a high degree of sequence identity may define the core structures of calponin family proteins (Figure 8).

The structural conservation of transgelin and calponin is not restricted to the CH domain that is found in various functionally diverse proteins (Gimona et al., 2002). The also conserved structures of the C-terminal portion of transgelin-1/SM22α corresponding to the middle region of calponin contains the actin-binding and phosphorylation regulatory sites (R. Liu & Jin, 2016a), forming the basis for their potentially conserved functions and regulations. The C-terminal segment of transgelin-1/SM22α modulated by low tension-induced proteolytic cleavage indicates a regulatory mechanism with predicted functional importance (R. Liu et al., 2017).

The sequence similarity between transgelin and the conserved N-terminal and middle regions of calponin suggests not only a common evolutionary ancestry but also a notion that transgelin represents a "mini-calponin" without the evolutionarily added C-terminal variable region seen in the calponin isoforms. With the ancestral position of *TAGLN2* among the vertebrate transgelin isoform genes (Figure 6), transgelin-2 may represent an ancestor-like prototype of the calponin family proteins, providing an informative experimental system to understand the core functions as well as the structure-function relationships of calponin.

In summary, while the three calponin isoforms mainly differ in their C-terminal structures the transgelins lack a C-terminal variable region (R. Liu et al., 2017). Transgelin and calponin both contain a highly conserved CH domain (M. Li et al., 2008). Transgelin contains only one CLIK motif versus the three found in calponins. While calponin has two actin-binding sites, transgelin has only one (Y. Fu et al., 2000). Phylogenetic lineage analysis demonstrated that calponin and transgelin diverged early during vertebrate evolution (Figure 9) (T.-B. Hsieh & Jin, 2023). As transgelin shows high sequence homology to calponin but is without structural repeats or the variable region, it is likely that transgelin represents an ancestral structure of calponin-transgelin family proteins. This "abbreviated" form conserves the essential core functions of the calponin-transgelin family of regulatory proteins while notably structural and functional divergences have occurred in the calponin isoforms through evolutionary selection.

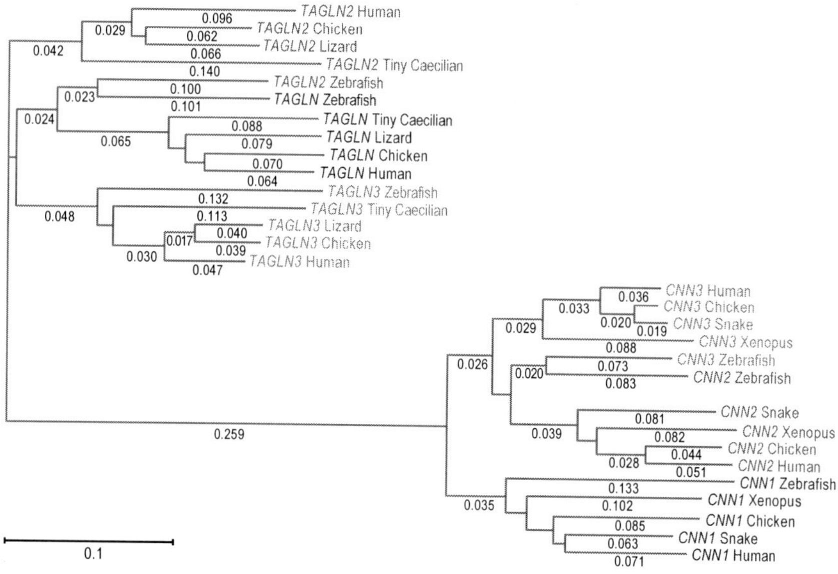

Figure 9. Phylogeny of the calponin-transgelin superfamily genes in vertebrates. The combined phylogenetic tree of vertebrate calponin and transgelin isoforms was generated by aligning amino acid sequences of representative species with the DNASTAR MegAlign computer software (Lasergene, lnc, Madison, WI) using Clustal W method. The phylogenetic lineages demonstrate that calponin and transgelin diverged early during vertebrate evolution. The NCBI database accession numbers of the sequences are listed in the legends of Figure 6 and Figure 2 in Chapter 1.

Chapter 15

Calponin-Like and Transgelin-Like Proteins in Invertebrates

Homologs of vertebrate calponin and transgelin have been found in invertebrate species. A calponin-like protein was first identified in mussels as a regulatory protein with actin and tropomyosin-related cellular functions. It is larger than vertebrate calponins containing a CH domain and five CLIK repeats (Dobrzhanskaya et al., 2013; Sirenko et al., 2013). Several other calponin-like proteins have been reported in invertebrates containing a CH domain and various numbers of CLIK repeats ranging from 7 in tapeworm to 23 in mollusks (S. Ono, 2021). unc-87, a calponin-like protein, of *Caenorhabditis elegans* contains 7 CLIK repeats in the absence of the CH domain, possibly representing an ancestral form of calponin and transgelin family proteins (K. Ono et al., 2015; Yamashiro et al., 2007).

The structural variations of invertebrate calponin-like and transgelin-like proteins are significantly larger than that in their vertebrate homologs. To understand the function of the conserved core structure of calponin-transgelin family proteins as well as the significantly diverged structures of their isoforms is of biological and medical importance. To investigate the molecular evolution of the calponin-transgelin gene family in the context of the structure-function relationships of calponin and transgelin and their isoforms across invertebrate and vertebrate species may provide a foundation for better understanding the biological functions of calponin and transgelin in various cell types as regulators of physiological and pathological processes.

15.1. Broad Presence of Calponin-Like and Transgelin-Like Proteins in Invertebrate Animals and Single Cell Organisms

Homologs of calponin-transgelin family proteins are found in invertebrates, for example unc-87 in *Caenorhabditis elegans* (Goetinck & Waterston, 1994) and mp20 in *Drosophila melanogaster* (Assinder et al., 2009; Ayme-Southgate et al., 1989). While vertebrate muscle tissues are classified as skeletal, cardiac and smooth muscles, muscle tissues of invertebrates are grouped into transverse striated, oblique striated and smooth muscles with variations among species, and intermediate types between transversely and oblique striated muscled or between obliquely striated and smooth muscles (Plesch, 1977). Same as in vertebrates, calponin is not expressed in highly differentiated striated muscles of invertebrates whereas studies have found calponin expression in invertebrate muscle types structurally similar to vertebrate smooth muscle (Royuela et al., 1997). Calponin was found in the smooth and oblique striated muscles of Helix snail (Royuela et al., 2000) and arm muscles of Antarctic starfish (Meyer-Rochow et al., 2003). During the embryonic development of marine spoon worm *Echiura*, striated and smooth muscle specific genes including calponin-like protein are expressed in specific muscle layers of adult body wall (Y. H. Han et al., 2020). A study found that freshwater sponge *Ephydatia muelleri* has three transgelin paralogs of which a transgelin-2-like protein was located in contractile bundles in pinacocytes (Colgren & Nichols, 2022).

Schistosoma japonicum expresses a 38-kDa protein with substantial structural similarity to vertebrate calponin exclusively in smooth musculature (W. Yang et al., 1999). This calponin-like protein is located in smooth myofibrils of adult worm in association with myofilaments. It is also present in smooth muscles of the forebody and stratified muscle of the tail. This protein is involved in the contraction of stratified tail muscle, indicating a function similar to smooth muscle calponin in vertebrates. Calponin-like proteins have been found in regulating octopus muscular contraction and a calponin 2-like isoform was identified in *Octopus bimaculoides* with an interesting function to produce antibacterial peptides against Gram-positive and Gram-negative bacteria (Maselli et al., 2020). Another study found that calponin expression significantly increased in the honeycomb moth *Galleria mellonella* 24 hours after inoculation of candida pathogens, implicating a role of calponin in cellular responses to infections (Sheehan & Kavanagh, 2018).

15.2. Function and Regulation of Calponin-Like and Transgelin-Like Proteins in Invertebrates

Calponin-like and transgelin-like proteins have been found with various other functions in invertebrate organisms besides regulating smooth muscle contraction. During larval-pupal metamorphosis of the cotton bollworm

Helicoverpa armigera, calponin is expressed at a high level in the metamorphic epidermis and phosphorylated by PKC (Q. Fu et al., 2009). In the embryonic stage, the sea squirt *Ciona* expresses a calponin-transgelin family gene that is crucial to notochord development (Oonuma et al., 2021). It was shown that non-phosphorylated calponin interacts with ultraspiracles protein to activate the juvenile hormone pathway and antagonize the 20-hydroxyecdysone pathway in initiating insect molting and metamorphosis (Cai et al., 2014). Calponin-like proteins also function in invertebrates during adaptation to environmental changes. In the non-calcifying marine worm *Platynereis dumerilii*, expression of a calponin-like gene changed significantly in response to low pH environment (Wage et al., 2016).

15.3. Phylogeny of Vertebrate and Invertebrate Calponin-Transgelin Super Family Proteins

A phylogenetic analysis of invertebrate calponin-like proteins has recently been published (T.-B. Hsieh & Jin, 2023). As invertebrates include 97% of animal species with highly diverged biological backgrounds, the current sequence database indicates that invertebrate calponin is more diverged among species but less distinguishable between isoforms in comparison to that of the three vertebrate isoforms. For example, arthropods represent the most numerous species of animals with over 30 million species (Stork, 2018) but have calponin isoforms closely clustered together despite marine or terrestrial species. This pattern is also shown in the phyla of roundworms and flatworms while their morphological features, organ structures and living environments are very different.

Similar patterns are found in the phylogeny of invertebrate transgelin and their isoforms (T.-B. Hsieh & Jin, 2023). Transgelin-like proteins have been identified in fungi, for example *STG*1, a protein of 174 amino acids, in fission yeast (Nakano et al., 2005) and *SCP*1, a protein of 200 amino acids, in budding yeast (Goodman et al., 2003; Winder et al., 2003). The transgelin-like proteins appear evolutionarily conserved in the kingdom of fungi.

A combined phylogenetic analysis of calponin and transgelin isoform genes in vertebrate and invertebrate animals demonstrated the overall phylogeny of calponin-transgelin superfamily proteins (T.-B. Hsieh & Jin, 2023). The result show that while calponin and transgelin both emerged early during animal evolution indicating fundamental biological functions, transgelin likely has the ancestor-representing prototype structure and functions with fundamental housekeeping roles as reported in the literature (Jo et al., 2018; Meng et al., 2017). The early emergence of transgelin-like proteins in single cell organisms (T.-B. Hsieh & Jin, 2023) also supports the hypothesis that transgelin-like proteins may have a fundamental biological function as a prototype of the calponin-transgelin family proteins. Therefore, the core structural and functional domains of vertebrate transgelin (Figure 8) merit more detailed characterizations.

Chapter 16

Mechanoregulation of Calponin and Transgelin

Mechanical forces have significant effects on various biochemical and genetic processes in living organisms and contribute to multiple physiological regulations and pathological conditions (N. Wang et al., 2009). The actin cytoskeleton is a dynamic network in eukaryotic cells capable of bearing forces and undergoing rearrangements in adaptation to mechanical changes in the environment, playing essential functions in cellular mechanical properties and responses to mechanical signals (J. Kang et al., 2011). For the dual function of actin cytoskeleton in generating as well as sensing mechanical forces, regulation of actin cytoskeleton is important for the mechanoregulation in eukaryotic cells.

It is well established that chemical energy can be converted to mechanical force in biological systems by motor proteins such as the myosin ATPase, and that mechanical signals, including static force such as tissue stiffness and gravity, potently induce cellular biochemical responses. However, the mechanisms that cells sense and convert the mechanical force signals into biochemical signals are incompletely understood. In the past two decades, increasing amounts of experimental research data have laid a foundation to further investigate cellular mechanosignaling and regulation. While the mechanisms of cellular sensing of dynamic mechanical force via the function of membrane channels and receptors have been established in advance, particularly for the mechanosensitive ion channel protein *Piezo1* (Lai et al., 2022), the mechanisms by which cells sense static mechanical force signals such as gravity and tissue stiffness require more in depth investigations.

Calponin and transgelin, especially calponin 2, have been shown to regulate cytoskeleton-based cell motility functions under mechanoregulation (Rasmussen & Jin, 2024). The expression of *CNN2* and *TAGLN* genes and the turnover of calponin 2 and transgelin-1/SM22α proteins are both themselves sensitive to mechanoregulation (Hossain et al., 2006; Jiang et al., 2014; R. Liu et al., 2017). The final chapter of this book will discuss mechanoregulation and functions of the calponin-transgelin family proteins in cellular activities and responses to mechanical force signals in the context of physiological and pathological processes discussed in the previous chapters. The main objective is to summarize the current knowledge of the functions of calponin and transgelin in cellular mechanoregulation for further investigations into how cells sense static tension and convert the force signal into biochemical and cellular activities.

16.1. Mechanosensing in Cell Regulations

Cellular modulation and subsequent response to mechanical forces affect many biological processes. In order to understand the molecular mechanisms and key players in cellular mechanosensing, it is important to first define mechanosensing. Mechanosensing is the ability of cells to perceive mechanical force, with mechanotransduction then being the interplay of these sensed forces with cellular responses to initiating a conversion to biochemical signals. Through the actin cytoskeleton, aided by a number of modulatory proteins, including calponin and transgelin, a cell performs the dual roles of mechanosensing and mechanotransduction to interact with and respond to the extracellular environment.

16.1.1. Cellular Sensing of Dynamic/Varying Mechanical Forces

Cellular sensing of dynamic mechanical force via the function of membrane channels and receptors (Mukherjee et al., 2022) has been established in detail, represented by the mechanosensitive *Piezo* ion channel proteins. *Piezo1* is a widely distributed mechanosensitive non-selective cation channel implicated in key cardiovascular, lung, urinary, and immune functions. The signaling pathways activated by mechanical stimulation of *Piezo1*, the downstream targets of *Piezo1*, and the overall effects of *Piezo1* activation have been extensively studied (Lai et al., 2022). *Piezo1* can sense acute structural changes of the plasma membrane caused by mechanical force applied to the cell. Growing evidence has shown that the actin cytoskeleton services as a converging point of various signaling pathways modulating the activity of ion transport proteins in cell membrane including *Piezo* channels in the sensing of acute mechanical force signal applied to plasma membrane (Morachevskaya & Sudarikova, 2021). On the other hand, the role of *Piezo1* in sensing static forces such as gravity is only via an indirect mechanism (J. H. Lee et al., 2024).

16.1.2. Mechanosensing of Static Force Signals

The actomyosin cytoskeleton is a major contractile and tension bearing cellular sensor system. In the "tethered model" (De et al., 2010; Fedorchak et al., 2014; Ingber, 1999; C. G. Lim et al., 2018; Yap et al., 2018), the actomyosin cytoskeleton is tethered to cell-cell or cell-extracellular matrix contacts that act to sense force and relay this to the actin filaments, which often mechanotransduces these signals through direct physical tethering to the nucleus where cellular gene expression can then be altered. The proteins that act as these tethering mechanosensors are an expanding list, among them are ion channels such as the *Piezo* channels, cytoskeletal proteins such as α-catenin and filamin, adhesion receptors such as PECAM-1 and cadherins, and extracellular ligands such as TGF-β (C. G. Lim et al., 2018).

The actin cytoskeleton acts as the primary structural support in most cell types, and the actin filaments are highly dynamic, with the ability to rapidly assemble and disassemble to assist the cell in sensing of shear force, gravity, and tissue stiffness, as well as in forming protrusions that promote movement. In addition, cells also have intermediate filaments that play a more passive reinforcement role, and microtubules that have a role in resisting compressive forces and intracellular transport (De et al., 2010).

The formation of actin stress fibers, bundles of actomyosin and cross-linking proteins such as α-actinin, are another important aspect of cellular mechanosensing. These complexes typically form within minutes of cell spreading and are largely mediated through activation of the GTPase RhoA. RhoA appears especially important for rapid responses to changes in cellular stress and actin cytoskeleton remodeling, with downstream effectors including Rho-associated kinase (ROCK) which inhibits myosin light chain phosphatase, and mDia which promotes actin polymerization (S. M. Lim et al., 2012; Matthews et al., 2006; Takemoto et al., 2015). More recent discoveries have shown that RhoA signaling also acts as an intermediary between the actin cytoskeleton (largely bound to adherens junctions through E-cadherin) and intermediate filaments coupled to desmosomes (Nanavati et al., 2023).

A mechanical stress signal is then typically relayed to the nucleus to initiate biochemical signals. This is largely through the Linker of Nucleoskeleton and Cytoskeleton (LINC) complex, which results in changes in chromatin conformation, gene activation and protein expression to alter cellular structure and function (Alam et al., 2016; Fedorchak et al., 2014; Pennacchio et al., 2021). YAP/TAZ play a role in delivering mechanical signal to nucleus transcriptional machinery in the HIPPO pathway (Dobrokhotov et al., 2018). In contrast to the sensing of dynamic force signals, the cellular sensing of static force signals is much less understood. The cytoskeleton tension-dependent expression of calponin and transgelin, and their functions in regulating myosin motor activity and actin cytoskeleton tension, render them attractive objectives for studies of cellular mechanosensing.

16.2. Calponin 2 Regulates Cell Traction Force

Cell traction force (CTF) generated by cytoskeleton myosin motors plays a critical role in controlling cell shape, enabling cell motility, and maintaining cellular mechanohomeostasis in many cell motility-related processes such as angiogenesis, embryonic and postnatal development, wound healing, and cancer metastasis. An established biochemical function of calponin is the inhibition of myosin ATPase. Calponin 2 is expressed in epithelial cells, endothelial cells, macrophages, myoblasts and fibroblasts, and plays a role in regulating cytoskeleton activities such as cell adhesion, migration and cytokinesis. Knockout of *CNN2* gene that encodes calponin 2 increased cell motility, suggesting a function of calponin 2 in modulating CTF. A CTF microscopy study measured the root-mean square traction, total strain energy, and net contractile movement of primary fibroblasts isolated from *CNN2* KO mice cultured on polyacrylamide gel substrates embedded with fluorescent beads (Hossain et al., 2016). The results showed that calponin 2-null fibroblasts exhibit greater traction force than that of WT cells.

Adherent calponin 2-null fibroblasts de-adhered faster than WT control during mild trypsin treatment, consistent with increased CTF. Mild treatment with blebbistatin, an inhibitor of myosin II ATPase (Allingham et al., 2005), which does not cause drastic change in morphology, produced a faster and stronger effect on altering cell morphology when calponin 2 is present in WT fibroblasts than that on *CNN2* KO cells. The dose-dependence and time course of blebbistatin's effect on mouse fibroblasts in adherent culture showed that *CNN2* KO fibroblasts were significantly less sensitivity to blebbistatin than that of WT cells, indicating a less endogenous inhibition of the force generated by myosin II motors when calponin 2 is absent. The effect of blebbistatin on diminishing the difference in CTF between *CNN2* KO and WT fibroblasts demonstrate that calponin 2 regulates CTF via inhibiting myosin II ATPase and motor activity (Hossain et al., 2016).

The effects of calponin 2 and blebbistatin are additive, indicating parallel mechanisms where calponin inhibits the activation of myosin by actin and blebbistatin directly inhibits myosin ATPase. Therefore, their additive inhibitory effects on CTF have two different components: The effect of calponin 2 is an endogenous regulatory inhibition to lower the degree of myosin II motor activation, whereas blebbistatin is a toxin that kills myosin ATPase and motor activity. This difference is critical for calponin 2 to function as a physiological modulator of cell traction force and motility in contrast to the effect of blebbistatin on paralyzing all myosin motor-based functions. The finding that calponin 2 regulates myosin-dependent CTF in non-muscle cells demonstrates its role in the mechanoregulation of cell motility-based functions (Hossain et al., 2016).

16.3. Mechanoregulation of Calponin and Transgelin

Much of the current research of calponin-transgelin family proteins in mechanosensing has focused on the function and expression of calponin 2 in response to mechanoregulation, a chicken-and-egg paradigm whereby which calponin 2 expression changes with changing tension and in turn regulates this tensile force production. The association of calponin 2 with the actin cytoskeleton is a foundation for its modulatory role in cellular mechanoregulation (Hossain et al., 2005). Although less studied, calponin 3 is also capable of regulating cytoskeleton mechanical tension and appears to be especially crucial for the activation of stress fibers (Daimon et al., 2013).

In the tethering model described above where the actin filaments have one end tethered to cellular junctions and another to the nucleus, calponin may act as a mechanoregulator by adjusting the dynamics and intrinsic tension of the actin string through regulating myosin ATPase (Abe et al., 1990; D. C. Tang et al., 1996) to modulate the sensing of static force signals. Whether transgelin also plays a similar role in this model remains to be investigated. Studies have shown that increased stretch applied to vascular smooth muscle tissue activates RhoA and increases synthesis of calponin and transgelin, but more research is needed to provide a clear linkage between these two pathways (Hellstrand & Albinsson, 2005; L. Liu et al., 2019).

16.3.1. Transcriptional Regulation

Culturing cells on substrates of different stiffness is an effective approach to quantitatively alter the mechanical environment and study the cellular responses. The transcriptional activity of mouse *CNN2* gene promoter-reporter constructs examined in transfected NIH/3T3 fibroblasts, HEK293 epithelial cells and C2C12 myoblasts showed responsiveness to the stiffness of polyacrylamide gel substrate. The segment between -1.57 kb and -1.38 kb in the 5'-region of mouse *CNN2* promoter contains a low substrate stiffness-induced repressor corresponding to an HES-1 (hairy and enhancer of split-1) transcription factor binding site (Figure 10A), where an increase of HES-1 in cells cultured non-adherently (experiencing a low stiffness environment) down-regulated the expression of calponin 2 (Jiang et al., 2014). HES-1 has been observed to function downstream of the Notch-RBP J signaling pathway (Kageyama et al., 1997) which has been suggested to mediate cellular mechanoregulations (J. Chen & Zolkiewska, 2011; Morrow et al., 2005, 2007; Teixeira et al., 2009).

Further experiments showed that treatment of cells cultured on soft gel substrate with Notch inhibitor DAPT increased the level of calponin 2 expression by diminishing the low tension-induced repression of *CNN2* gene. Consistently, Notch activator PEITC significantly decreased the level of calponin 2 expression. In contrast, DAPT and PEITC had minimum effects on the expression of calponin 2 in adherent monolayer cells growing on plastic substrate, suggesting that the higher mechanical tension built in cells adhered to high stiffness matrix has a dominant effect overriding the effect of Notch signaling (Jiang et al., 2014). The results suggest that Notch signaling plays a role in the regulation of *CNN2* gene expression (Figure 10B), which is dependent on mechanical tension in the cytoskeleton built against the stiffness of extracellular environment (Hossain et al., 2005, 2006). The key finding that high stiffness substrate-generated cytoskeletal static tension dominantly overrides the downregulating effect of Notch activator on calponin 2 expression demonstrates the potency of mechanosignaling in regulating cellular functions (Jiang et al., 2014).

In contrast, transfective expression of calponin 2 under a viral promoter that is not regulated by the intrinsic cellular mechanoregulation showed no decrease of calponin 2 protein when the cells were cultured on soft versus hard gel substrate (Hossain et al., 2006). This observation indicates that transcriptional regulation is a primary determinant for the cellular response to mechanical signals, at least in the case of calponin 2.

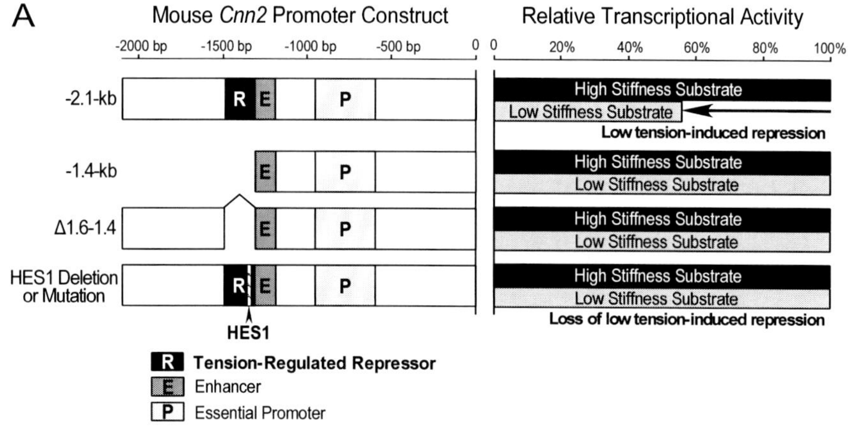

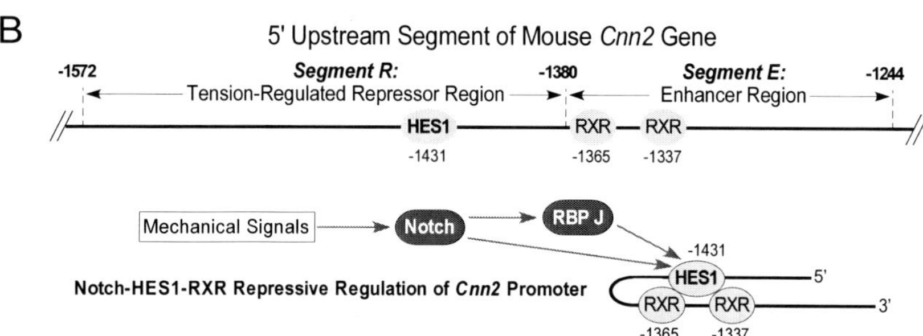

Figure 10. Notch-HES1-mediated mechanoregulation of *CNN2* promoter. A. Using serial and internal deletion and HES1 site mutation promoter-reporter constructs, the transcriptional activity of *CNN2* gene promoter was mapped in cells cultured on high versus low stiffness gel substrates. The essential promoter activity (P) located in the -1002 to -611 region is required to express the *CNN2* gene. The -1380 to -1244 region contains enhancer activity (E) required for high level of expressions. Functions of both of these regions are independent of the stiffness of culture substrate. A low tension-induced repressor activity (R) was identified in the -1572 to -1380 region, which is based on the function of an HES-1 site. B. Mechanoregulation-related cis-transcriptional regulatory elements in mouse *CNN2* promoter were predicted using the MatInspector software. Two RXR binding sites are found in the enhancer segment of -1380 to -1244 downstream of the HES-1 site. A model is proposed for a potential mechanism of Notch-RBP J-HES1-mediated mechanoregulation of the *CNN2* promoter activities (Jiang et al., 2014).

Several other cell signaling pathways have been suggested for functions in cytoskeleton-based mechanoregulation of eukaryotic cells. For example, the angiotensin II receptor related pathway (de Cavanagh et al., 2009), the integrin-mediated pathway (J. Li et al., 2008), and the G protein-coupled receptor pathway (Tzima, 2006). The effect of reduction of cytoskeleton tension by blebbistatin inhibition of myosin motor activity on decreasing calponin 2 expression is without decreasing the concentration of actin fibers, indicating that calponin 2 gene expression is directly responsive to changes in cytoskeletal tension (Hossain et al., 2006). The interrelationship between calponin regulation of myosin motor force production and Notch-mediated mechanoregulation is worth further investigation.

The expression of transgelin is also regulated by mechanical tension, shown by studies performed in vascular smooth muscle where increased stiffness increases transgelin-1/SM22α expression detected at both mRNA and protein levels (R. Liu et al., 2017). Collagen and Notch signaling may also play regulatory roles, where type IV collagen was shown to inhibit the expression of both calponin and transgelin genes involving Notch signaling (X. Zhang et al., 2013). Studies also showed that transgelin-2 participates in the progression of colorectal cancer through interacting with CD44 to regulate the Notch1 signaling pathway (Ding et al., 2022). Transgelin's association with the actin-myosin filaments in the cytoskeleton (Shapland et al., 1993) provides a structural basis for transgelin to sense and respond to the mechanical tension generated by myosin motor in the actin cytoskeleton, and suggests an interrelationship with calponin 2. These findings link functional and mechanistic similarities between the mechanoregulation and function of calponin and transgelin. Further studies are merited to explore whether the low tension-dependent repressor mechanism that regulates *CNN2* gene expression (Jiang et al., 2014) is also present in the promoter of *TAGLN* and related genes.

16.3.2. Posttranslational Mechanoregulation

Rapid degradation of calponin 2 occurs in alveolar cells after the reduction of mechanical tension in collapsed lung tissue, which can be effectively prevented by maintaining postmortem mouse lung at the in vivo inflated volume (Hossain et al., 2006). This low cytoskeletal tension-induced degradation of calponin 2 was further confirmed in cells cultured on continuously stretched silicon rubber membrane for 3 days and then releasing the tension (Hossain et al., 2006). Similarly, a degradation of transgelin protein occurs with a reduction of static mechanical tension of aortic rings in culture (R. Liu et al., 2017). Considering that calponin and transgelin are abundant cytoskeleton proteins, the coordinated mechanoregulations of their levels transcriptionally and post-translationally convey effective and rapid cellular responses to changes in the mechanical environment.

Mediated by PKC phosphorylation, a rapid degradation of calponin 2 also occurs during metaphase of the cell cycle and is required prior to cell division (Qian et al., 2021). Cytokinesis is a dynamic mechanical process in which a decrease in cytoskeletal tension, as proposed in the polar-relaxation model (Reichl et al., 2005), may contribute to the rapid degradation of calponin 2 to facilitate the remodeling of cellular structure and cleavage by the F-actin contractile ring (Qian et al., 2021).

The mechanoregulation of *CNN2* gene expression is primarily a low tension-induced repression of transcription (Jiang et al., 2014). Therefore, a posttranslational down-regulation of calponin 2 in the actin cytoskeleton by low tension-induced proteolysis (Hossain et al., 2006) would decrease F-actin stability and in turn decrease the tension signal transduced into and sensed by the cell, resulting in a repression of gene transcription to maintain a low level equilibrium of calponin 2 in the cell. On the other hand, expression of calponin 2 under viral promoter that lacks such mechanoregulation showed no decrease but a significant increase in calponin 2 protein when the cells were cultured on soft versus hard gel substrate (Hossain et al., 2006). Therefore, low cytoskeleton tension may actually promote accumulation of calponin 2 protein in the cell, thus the downregulation at transcriptional level is critical to the effectiveness of cellular response to mechanical signals.

While the mechanoregulation of transgelin-1/SM22α by protein degradation and gene expression resembles that of calponin 2, the expression of transgelin-1/SM22α responds to remarkably higher matrix stiffness than that of calponin 2 in the same cell type. *CNN2* KO significantly decreases the overall expression and tension sensitivity of transgelin-1/SM22α. With their evolutionary homology, structural similarity and cytoskeleton co-localization, transgelin-1/SM22α and calponin 2 may have interrelated regulations and functions in the mechanoregulation of actin cytoskeleton and cell motility.

16.4. Role of Calponin and Transgelin in Cell Mechanoregulation

Calponin has been shown to exert transcriptional control over cellular signaling pathways, including the PI3K/Akt signaling pathway, where overexpression of *CNN3* in non-small cell lung cancer lines suppressed

PI3K/Akt expression (C. Yang et al., 2021). This pathway is well known to affect cell proliferation and is highjacked by many cancers to promote aberrant growth and metastasis. In addition, *CNN2* knockdown increased PI3K/Akt and NF-kB expression and decrease E-cadherin expression in pancreatic ductal adenocarcinoma cells (Qiu et al., 2017), indicating that calponin 2 and calponin 3 exert similar transcriptional effects on PI3K/Akt and NF-kB mediated cell signaling.

However, *CNN3* knockdown in proliferating myoblasts induced an opposite effect on inhibiting the PI3K/Akt pathway as measured by a decrease in phosphorylated Akt and decreased mTOR, a downstream target of Akt (She et al., 2021). Similarly, transgelin also has been shown to interact with the PI3K/Akt pathway with overexpression of transgelin activating this pathway (L. Liu et al., 2019). These observations suggest that calponin and transgelin may function as a balancing modulator in mechanoregulation to adjust cellular activities at various differentiation and functional states.

The nature of calponin and transgelin as cytoskeleton mechanical regulators, of which their gene expression and protein turnover are also regulated by mechanical tension, places them at a unique position for functions in cellular mechanical sensing and responses. Further investigation is needed to establish their functional mechanisms. The Notch-HES1 regulation of *CNN2* gene expression (Jiang et al., 2014) and PKC-phosphorylation-induced dissociation of calponin from F-actin (J. P. Jin et al., 2000) and rapid degradation (Qian et al., 2021) of calponin 2 protein provide leads for future mechanistic studies.

16.5. Binding Partners of Calponin with Potential Mechanisms of Sensing and Transduction of Mechanical Signals

Calponin acts in combination with many known binding partners with potential roles in accomplishing mechanoregulation in cells. The structural domains of calponin are defined by both their function and ability to interact with these binding partners (see Figure 4 in Chapter 2) (T.-B. Hsieh & Jin, 2023; R. Liu & Jin, 2016a). While the binding partners of calponin have largely been discovered through co-precipitation studies, which run the risk of identifying proteins that instead co-localize on actin filaments, several of them are relevant to cellular mechanoregulation with evidence of functional effects of the interactions.

16.5.1. Calmodulin

Calmodulin is a small, calcium-binding protein containing four EF-hand ion-binding motifs and is abundant in the cytoplasm, where it acts as an intermediary messenger (Vetter & Leclerc, 2003). Calmodulin has been shown binding to calponin (reversable by Ca^{2+}) and inhibits the phosphorylation of calponin by PKC in a noncompetitive fashion, thus regulating the binding of calponin to F-actin (Naka et al., 1990). The calmodulin-binding site is located in the N-terminal domain of calponin, distant from the PKC phosphorylation site Ser_{175} (Winder & Walsh, 1993). Other protein binding studies have shown that calmodulin has a weak affinity for calponin, and it has been argued that non-physiologically-relevant high concentrations of calmodulin are required for in vitro studies to produce calmodulin's regulation of calponin phosphorylation (Wills et al., 1993; Winder, Allen, et al., 1993). Despite this potential limitation, the possibility that calponin provides a crucial link between cellular calcium signaling and the subsequent actin cytoskeleton mechanotransduction merits continued investigation.

16.5.2. Caldesmon

Caldesmon is an actin- and calmodulin-binding protein which also binds myosin to inhibit myosin ATPase (Lin et al., 2009), through which it is involved in regulating smooth muscle contraction. Calponin and caldesmon show strong binding interactions (Graceffa et al., 1996) implicating interrelated functions in the regulation of smooth muscle contractility (Huber, 1997). Actin filaments contain both contractile and cytoskeletal domains in smooth muscle cells. Unlike calponin, which localizes to both the contractile and cytoskeletal domains of actin, caldesmon is found only in the contractile domain of smooth muscle and does not relocate upon stimulation of contraction (Graceffa et al., 1996; Khalil et al., 1995; Parker et al., 1994). However, caldesmon also has a non-muscle alternative splice form that plays a role in regulating functions of the actin cytoskeleton (Lin et al., 2009). Calponin and caldesmon can bind F-actin simultaneously, with calponin binding outcompeting caldesmon (Makuch et al., 1991). The current view is that calponin and caldesmon may perform complementary and largely mutually exclusive functions in regulating actin filaments, with calponin more predominantly in the cytoskeletal domain and caldesmon more predominately in the contractile domain (Gusev, 2001). As caldesmon is a protein that couples actin filaments with myosin motors (Lin et al., 2009), its contribution to cellular mechanics and the significance of its interaction with calponin in cellular mechanical homeostasis deserve more in depth investigation.

16.5.3. PKC and ERK1/2

The family of PKC, a serine/threonine kinase, is a group of ubiquitous calcium-dependent cellular regulators. In smooth muscle, PKC has been shown to inhibit myosin light chain phosphatase, thus enhancing smooth muscle contraction (Ringvold & Khalil, 2017). Calponin 1 has been shown to co-precipitate with PKC and ERK1 and produce PKC phosphorylation possibly by promoting autophosphorylation that may be necessary for the activation of PKC in smooth muscle (H. R. Kim et al., 2013; B. Leinweber et al., 2000). PKC has also been shown to directly phosphorylate calponin (Andrea & Walsh, 1992). Further studies have shown that calponin 3 has similar functions, suggesting the likelihood of action through a conserved functional domain (Appel et al., 2010). In the meantime, PKC phosphorylation-induced rapid degradation of calponin 2 is required for normal completion of cytokinesis (Qian et al., 2021). With these evidences, it has been theorized that PKC may act through a calponin/caldesmon-dependent kinase cascade involving ERK kinase to regulate smooth muscle contraction, although the exact steps in this process remain unclear (Je et al., 2001; H. R. Kim et al., 2013). Strain force rapidly activated both ERK1/2 and p38 in smooth muscle (Guest et al., 2006), therefore, the PKC and ERK1/2 signaling pathway in relationship with calponin's role in mechanosensing and regulation of cellular responses needs to be further studied.

16.6. Research Directions and Hypothesis

As a family of actin cytoskeleton regulatory proteins that are both ubiquitously expressed in many cell types and uniquely expressed in high levels in smooth muscle and migrating and tension-bearing cells, calponin and transgelin represent attractive targets for understanding many physiological and pathological mechanosignaling processes. Among these include blood pressure regulation, blood coagulation, renal glomerular filtration, ovarian follicle development, inflammation, fibrosis, wound healing, and tumor metastasis. Studies have demonstrated that calponin 2 regulates various cytoskeleton dynamics-based cell motility functions under mechanoregulation. In addition to the examples outlined in the previous chapters of this book, deletion of calponin 2 increases phagocytosis of macrophages (Huang et al., 2008; R. Liu & Jin, 2016b) and degradation of calponin 2 is required in lysophagy (Kravić et al., 2022). Reflecting the underlying mechanisms, *CNN2* expression depends on the stiffness of extracellular matrix corresponding to the mechanical force built in the cytoskeleton by myosin motors that are conversely under an inhibitory regulation by calponin (Hossain et al., 2005).

The essential role of mechanosignaling in regulating the function of living cells is universally observed. It is important to understand how mechanical signals are transduced in cells to regulate gene expression and protein functions. The mechanisms that sense and transmit static mechanical force signals such as tissue stiffness and gravity into molecular and biochemical signals remain to be investigated. Chronic cyclic stretching of lung alveolar cells in culture did not increase but decreased the expression of calponin 2 (Hossain et al., 2006). This intriguing observation led to a hypothesis that after the cytoskeleton structure first remodeled to fit the longest dimension at the peak of stretches (to avoid structurally breakdown), the tension in the cytoskeleton will be decreased during the relaxation phases of the cyclic stretching program, resulting in a decrease, rather than increase, of the average cytoskeleton tension over time and subsequently decreasing the expression of calponin 2. This hypothesis is experimentally supported by quantitative data from cyclic stretching culture of fibroblasts under high, medium and low ratios of relaxed versus stretched durations, where short stretch (10% of the cycle duration, which still requires cytoskeleton remodeling to match the longest dimension as the reference point) and long relaxation (90% of the cycle) produced the lowest expression of calponin 2 (Figure 11) (Hossain et al., 2006; Rasmussen & Jin, 2024). This average tension overtime model supports the regulation of cellular activities by static mechanical force signals through the acumination of biochemical molecules over time. The quantitative responsiveness of *CNN2* gene expression to mechanical force signals as observed in multiple cell types studied to date provides an informative experimental system to investigate how cells sense static tension and convert the force signals to biological activities.

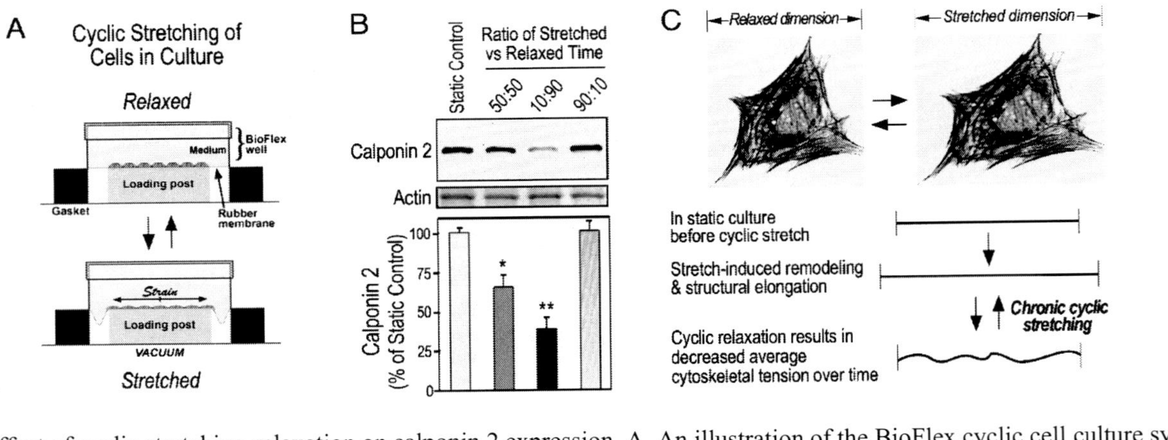

Figure 11. Effect of cyclic stretching-relaxation on calponin 2 expression. A. An illustration of the BioFlex cyclic cell culture system. B. Western blot and densitometry quantification of calponin 2 expression in NIH/3T3 fibroblasts cultured on the rubber membrane of BioFlex 3000C plates under cyclic stretching and relaxation for 3 days showed decreased levels. The degree of downregulation of calponin 2 expression shows a positive correlation to the percent of relaxation time where the 90% relaxation formula (2 sec stretch – 18 sec relaxation) produced more decrease than that produced by the 50% relaxation formula (2 sec stretch – 2 sec relaxation) in comparison to that in the static control cells cultured on rubber membrane that represents a high stiffness substrate to exert near maximum tension in the cytoskeleton while the 10% relaxation formula (18 sec stretch – 2 sec relaxation) did not produce a detectable change ($*p < 0.05$, $**p < 0.01$) (Rasmussen & Jin, 2024). The data demonstrate a regulation of calponin 2 expression by the average cytoskeleton tension over time. C. A model in which cells cultured on high stiffness substrates sense mechanical tension in reference to the steady cytoskeleton tension state. The two hypothetical actin-calponin cytoskeleton images of a cell represent the relaxed and stretched states during chronic cyclic stretching-relaxation. After the remodeling of cytoskeleton to match the stretched larger dimension, the cellular structures will become slack and experience less tension during the relaxed phase in reference to the near maximum tension equal to that in cells cultured statically. Therefore, data from chronic cyclic stretching cell cultures or organs undergoing cyclic dimensional changes in vivo, such as the lung, heart and arterial vessels, can be interpreted under a cyclic relaxation paradigm.

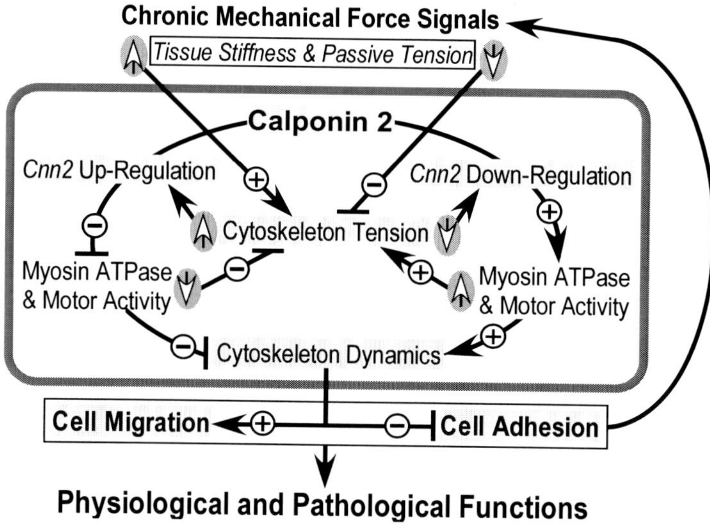

Figure 12. Calponin 2 in cellular mechanoregulation. Based on calponin's primary function as an inhibitory regulator of myosin motor to decrease the mechanical tension built in the cytoskeleton, a hypothetic model is proposed. Increases in *CNN2* gene expression will increase the inhibition of myosin motors and decrease cytoskeleton tension, which generates negative feedback to downregulate *CNN2* expression and maintain an equilibrium. *CNN2* downregulation will produce a series of opposite effects. The functions of calponin 2 and its tension-regulated expression contribute to the determination of a cellular tension set point that sustains a cell type-specific mechanical homeostasis critical to the biological functions of the cell. Extracellular mechanical force signals, such as tissue stiffness and passive tension, interplay with this regulation and impart physiological or pathological effects on functions such as cell adhesion and migration.

Living cells dynamically change structure and function through gene regulation and posttranslational protein modification in response to mechanical signals (Choquet et al., 1997; Eckes & Krieg, 2004; Hamill & Martinac, 2001; Lehoux & Tedgui, 2003; J. L. Walker et al., 2005). The cellular and systemic functions of calponin 2 are primarily based on its inhibitory regulation of actin cytoskeleton-associated myosin motors (Hossain et al., 2016). The interplay between the cytoskeleton and the mechanical environment has a potential role during cell motility-based activities. A hypothesis is proposed (Rasmussen & Jin, 2024) that the calponin-transgelin family proteins play a role in the equilibrium between cytoskeleton tension and the mechanical environment in which the cell resides. In this model, each cell type and functional state of a living cell have evolved or adapted with an intrinsic tension set point. When the set point or the mechanical environment changes, an imbalance of this equilibrium imparts structural and functional changes in the cell with physiological and pathological effects. Substrate stiffness and other extracellular mechanical force signals will influence the set point of this equilibrium depending on the cell type and their native tissue environments.

The specifically regulated expression and level of calponin 2 in different cell types may serve as a regulatory factor reflecting as well as determining the cellular tension set point. Changes in calponin level and/or function may be targeted to steer the cellular tension set point and adjust cell adaptation and response to the mechanical environment with physiological and pathological significances (Figure 12). Continuing research to establish the role of calponin and transgelin in cellular sensing and response to mechanical force signals and the underlying molecular mechanisms will help to better understand the mechanosensing and mechanoregulation in living systems and their interaction with the mechanical environment.

The mechanoregulation of *CNN2* gene expression and the function of calponin 2 in cytoskeleton stability and dynamics indicate broad physiological and pathological significance. Tissues and cells originated from all three embryonic germ layers have been found to express calponin 2. Various immortalized cell lines including epithelial, fibroblasts, myeloid, myoblasts and cancer express high levels of calponin 2 under the regulation of cytoskeleton tension (Rasmussen & Jin, 2024). Based on anatomical and functional features, these naturally calponin 2-positive cell types can be placed into three classes: the first is cells bearing constant or dynamic mechanical tension, such as epithelial and endothelial cells and smooth muscle cells in the wall of various internal organs; the second is migrating cells such as fibroblasts and macrophages; and the third group is proliferating cells including myoblasts and other stem cells. This classification may help to understand the physiological roles of calponin 2 as a mechanoregulatory molecule in different cell types.

It has been long established that chemical energy can readily be converted to mechanical force by ATPase- or GTPase-based motor proteins. It is also widely observed that mechanical force can signal cellular functions and gene expression. To further understand how the mechanical force energy is converted to chemical signals that trigger biological responses in a living cell is of major biological and medical importance. The mechanically regulated *CNN2* gene promoter provides not only an example of such signal transduction but also a powerful experimental system to search for the upstream cellular sensor of static mechanical signals.

Conclusion and Perspectives

Nearly four decades after its discovery, calponin is now known from extensive studies as an actin filament-associated regulatory protein with functions in modulating smooth muscle contractility and non-muscle cell motility. The three isoforms of calponin encoded by three homologous genes in vertebrates, *CNN1*, *CNN2* and *CNN3*, are of conserved structures in the N-terminal and middle regions whereas diverged in the C-terminal segment corresponding to significant size and charge differences. The three calponin isoform genes are expressed under tissue- and cell type-specific regulation, reflecting diverged biological functions. Calponin 1 is specifically expressed at abundance in smooth muscle cells and plays a role in modulating contractility. Calponin 2 is expressed in both smooth muscle and many types of non-muscle cells and functions as a regulator of the actin cytoskeleton in fundamental activities such as cell proliferation, migration, adhesion, phagocytosis, angiogenesis, immune responses, fibrosis and cancer metastasis. Calponin 3 is expressed in smooth muscle and several types of non-muscle cells with demonstrated functions in regulating neuronal plasticity, and cell fusion with a critical role in embryonic development.

CNN3 KO in mice results in impaired embryonic development and neonatal lethality. A direct fitness value of calponin 2 is demonstrated by the finding that its loss causes ovarian insufficiency and premature failure. It has also been observed that *CNN1*,*CNN2* double KO female mice have dystocia (our unpublished observation). The roles of calponin 1 and calponin 2 in maintaining smooth muscle contractility and blood pressure, inflammatory responses and blood clotting are also critical biological functions whereas calponin 1 and calponin 2 deletions can lower blood pressure, and *CNN2* KO slows the unset of thrombosis, alleviates inflammatory arthritis, reduces atherosclerosis, aortic valve calcification and post-operational peritoneal adhesion with potential for the development of new therapeutic approaches.

Calponin 2 is expressed at significant levels in three groups of cells: a) cells that are naturally under mechanical tension, b) physically mobile cells and c) rapidly dividing cells. Calponin 2 reduces myosin motor activity to stabilize actin cytoskeleton, which regulates the rate of cytokinesis, enhances cell adhesion, inhibits cell migration, affects secretion related cellular activities, underlying its biological functions and the therapeutic effects of its deletion. The finding that the gene regulation and function of calponin 2 are both under mechanoregulation opened a new direction in calponin research leading to better understanding of cellular interactions with the mechanical environment and responses to force signals.

A study has reported the use of calponin 2 as a blood-based biomarker for the early diagnosis of breast cancer. Enzyme-linked immunosorbent assay using mAbs against calponin 2 to determine the serum level of calponin 2 in samples from patients with breast cancer and benign diseases of the breast, and samples from healthy females showed increased serum levels of calponin 2 in 10.5% of patients with breast cancer in contrast to the negative finding in all of the control groups (Debald et al., 2014). The potential value of calponin detection as biomarkers of diseases presents another new direction of research.

While less functional studies have been done for transgelin and its three isoforms, the currently available data show their functions similar to that of calponin (Table 1). It is also worth noting that calponin and transgelin are both regulated by cellular mechanoregulation. For its broad cellular distribution, more in depth research is merited to further establish the biological functions and cellular regulations of transgelin with potentially medical significance.

Table 1. Expression, function and regulation of calponin and transgelin

Protein	Expression	Function	Mechanism	Regulation
Calponin	Smooth muscle cells	Smooth muscle contractility	Inhibiting actin activated myosin motor	PKC phosphorylation
	Fibroblast, lymphocytes, macrophage	Proliferation, adhesion, migration	Stabilizing actin cytoskeleton, tension related mechanotension	PKC phosphorylation, tyrosin phosphorylation, MEKK1 phosphorylation, methylation
	Macrophages	Phagocytosis		
	Neural cells	Cell plasticity		
	Trophoblast, myoblast	Cell fusion		
	Prostate cancer cells, ovarian cancer cells	Cell proliferation, migration, adhesion, and metastasis		
	HeLa, HEK293T cells	Lysophagy		
	Inner ear hair cells	Cytoskeleton structure		
Transgelin	Smooth muscle cells	Smooth muscle contractions	Regulate actin cytoskeleton	PKC phosphorylation
	Fibroblast, lymphocytes, epithelial cells, neural cells	Proliferation, adhesion, migration, phagocytosis	Stablize actin, mechanotrgulation	PKC phosphorylation, Tyrosin phosphorylation, MEKK1 phosphorylation, methylation, NF-κB, TGF-β, ERK
	Brease cancer cells, prostate cencer cells, colorectal cancer cells	Cancer cell proliferation, invasion, and metastasis		

Based on the current knowledge, many interesting questions are ready to be addressed to further understand the regulation and structure-function relationship of calponin and transgelin. For example, how do they regulate the smooth muscle actin filaments in contraction and the actin cytoskeleton in non-muscle cell motility? While phosphorylation clearly regulates calponin function in test tubes, why phosphorylated calponin is not accumulated in living cells? Are the functions of different isoforms of calponin exchangeable or distinct? How do mechanical tension signals transmit in cells to regulate the expression and function of calponin and transgelin? Are there any genetic diseases caused by calponin or transgelin mutations? Would calponin blood test be widely applicable in cancer diagnosis? Could pharmacological approach be developed to reduce or increase the expression and/or function of calponin 2 in various cell types for use in medical therapies? Future work along these directions will produce exciting information to translate the physiological and pathological significances of calponin and transgelin into medical applications to improve human health. The studies and developments presented in this book provide readers with a concise database index to explore new research and applications and move the field forward.

Bibliography

Abe, M., Takahashi, K., & Hiwada, K. (1990). Effect of calponin on actin-activated myosin ATPase activity. *Journal of Biochemistry*, *108*(5), 835–838. https://doi.org/10.1093/oxfordjournals.jbchem.a123289.

Agassandian, C., Plantier, M., Fattoum, A., Represa, A., & Terrossian, E. (2000). Subcellular distribution of calponin and caldesmon in rat hippocampus. *Brain Research*, *887*(2), 444–449. https://doi.org/10.1016/s0006-8993(00)03030-4.

Aikawa, E., Nahrendorf, M., Sosnovik, D., Lok, V. M., Jaffer, F. A., & Aikawa, M. (2007). Multimodality molecular imaging identifies proteolytic and osteogenic activities in early aortic valve disease. *Circulation*, *115*(3), 377–386.

Alam, S. G., Zhang, Q., Prasad, N., Li, Y., Chamala, S., Kuchibhotla, R., Kc, B., Aggarwal, V., Shrestha, S., Jones, A. L., Levy, S. E., Roux, K. J., Nickerson, J. A., & Lele, T. P. (2016). The mammalian LINC complex regulates genome transcriptional responses to substrate rigidity. *Scientific Reports*, *6*, 1–11. https://doi.org/10.1038/srep38063.

Allen, B. G., & Walsh, M. P. (1994). The biochemical basis of the regulation of smooth-muscle contraction. *Trends Biochem Sci*, *19*, 362–368.

Allingham, J. S., Smith, R., & Rayment, I. (2005). The structural basis of blebbistatin inhibition and specificity for myosin II. *Nat. Struct. Mol. Biol*, *4*, 378–379.

Almendral, J. M., Santaren, J. F., & Perera, J. (1989). Expression, cloning and cDNA sequence of a fibroblast serum-regulated gene encoding a putative actin-associated protein (p27. *Exp Cell Res*, *181*, 518–530.

Andrea, J. E., & Walsh, M. P. (1992). Protein kinase C of smooth muscle. *Hypertension*, *20*(5), 585–595. https://doi.org/10.1161/01.HYP.20.5.585.

Annes, J. P., Munger, J. S., & Rifkin, D. B. (2003). Making sense of latent TGFbeta activation. *J Cell Sci*, *116*(Pt 2), 217–224.

Ansorge, M., Sapudom, J., Chkolnikov, M., Wilde, M., Anderegg, U., & Moller, S. (2017). Mimicking Paracrine TGFbeta1 Signals during Myofibroblast Differentiation in 3D Collagen Networks. *Sci Rep*, *7*(1), 5664. doi: 10.1038/s41598-017-05912-x.

Appel, S., Allen, P. G., Vetterkind, S., Jin, J. P., & Morgan, K. G. (2010). h3/Acidic calponin: An actin-binding protein that controls extracellular signal-regulated kinase 1/2 activity in nonmuscle cells. *Mol Biol Cell*, *21*(8), 1409–1422.

Applegate, D., Feng, W., Green, R. S., & Taubman, M. B. (1994). Cloning and expression of a novel acidic calponin isoform from rat aortic vascular smooth muscle. *Journal of Biological Chemistry*, *269*(14), 10683–10690. https://doi.org/10.1016/s0021-9258(17)34113-3.

Arjunon, S., Rathan, S., Jo, H., & Yoganathan, A. P. (2013). Aortic valve: Mechanical environment and mechanobiology. *Ann Biomed Eng*, *41*(7), 1331–1346.

Arung, W., Meurisse, M., & Detry, O. (2011). Pathophysiology and prevention of postoperative peritoneal adhesions. *World J Gastroenterol*, *17*(41), 4545–4553. https://doi.org/10.3748/wjg.v17.i41.4545.

Assady, S., Benzing, T., Kretzler, M., & Skorecki, K. L. (2019). Glomerular podocytes in kidney health and disease. *Lancet*, *393*, 856-858.

Assinder, S. J., Stanton, J. A., & Prasad, P. D. (2009). Transgelin: An actin-binding protein and tumour suppressor. *Int J Biochem Cell Biol*, *41*, 482–486.

Ayme-Southgate, A., Lasko, P., & French, C. (1989). Characterization of the gene for mp20: A Drosophila muscle protein that is not found in asynchronous oscillatory flight muscle. *J Cell Biol*, *108*, 521–531.

Aziminia, N., Nitsche, C., Mravljak, R., Bennett, J., Thornton, G. D., & Treibel, T. A. (2023). Heart failure and excess mortality after aortic valve replacement in aortic stenosis. *Expert Rev Cardiovasc Ther*, *21*(3), 193–210.

Baig, A., Bao, X., & Haslam, R. J. (2009). Proteomic identification of pleckstrin-associated proteins in platelets: Possible interactions with actin. *Proteomics*, *9*, 4254-4258.

Bain, C. C., Hawley, C. A., Garner, H., Scott, C. L., Schridde, A., Steers, N. J., Mack, M., Joshi, A., Guilliams, M., Mowat, A. M., Geissmann, F., & Jenkins, S. J. (2016). Long-lived self-renewing bone marrow-derived macrophages displace embryo-derived cells to inhabit adult serous cavities. *Nat Commun*, *7*, *ncomms11852*. doi: 10.1038/ncomms11852.

Balachandran, K., Sucosky, P., & Yoganathan, A. P. (2011). Hemodynamics and mechanobiology of aortic valve inflammation and calcification. *Int J Inflam*, *2011*, 263870. doi: 10.4061/2011/263870.

Banno, Y., Nakashima, S., Hachiya, T., & Nozawa, Y. (1995). Endogenous cleavage of phospholipase C-beta 3 by agonist-induced activation of calpain in human platelets. *J Biol Chem*, *270*, 4318-4324.

Barany, M., & Barany, K. (1993). Calponin phosphorylation does not accompany contraction of various smooth muscles. *Biochim Biophys Acta*, *1179*, 229–233.

Bartegi, A., Roustan, C., Kassab, R., & Fattoum, A. (1999). Fluorescence studies of the carboxyl-terminal domain of smooth muscle calponin effects of F-actin and salts. *Eur J Biochem*, *262*, 335–341.

Baum, J., & Duffy, H. S. (2011). Fibroblasts and myofibroblasts: What are we talking about? *J Cardiovasc Pharmacol*, *57*(4), 376–379.

Bertazzo, S., & Gentleman, E. (2017). Aortic valve calcification: A bone of contention. *Eur Heart J*, *38*(16), 1189–1193.

Bitzer, M., & Wiggins, J. (2016). Aging Biology in the Kidney. *Adv Chronic Kidney Dis*, *23*, 12-18.

Blakney, A. K., Swartzlander, M. D., & Bryant, S. J. (2012). The effects of substrate stiffness on the in vitro activation of macrophages and in vivo host response to poly(ethylene glycol)-based hydrogels. *J Biomed Mater Res A*, *100*, 1375–1386.

Bogatcheva, N. V., & Gusev, N. B. (1995). Interaction of smooth muscle calponin with phospholipids. *FEBS Letters*, *371*(2), 123–126. https://doi.org/10.1016/0014-5793(95)00868-a.

Bondzie, P. A., Chen, H. A., Cao, M. Z., Tomolonis, J. A., He, F., Pollak, M. R., & Henderson, J. M. (2016). Non-muscle myosin-IIA is critical for podocyte f-actin organization, contractility, and attenuation of cell motility. *Cytoskeleton (Hoboken, 73*, 377-395.

Bostrom, K. I., Rajamannan, N. M., & Towler, D. A. (2011). The regulation of valvular and vascular sclerosis by osteogenic morphogens. *Circ Res*, *109*(5), 564–577.

Boyce, B. F., Schwarz, E. M., & Xing, L. (2006). Osteoclast precursors: Cytokine-stimulated immunomodulators of inflammatory bone disease. *Current opinion in rheumatology*, *18*, 427-432.

Bramham, J., Hodgkinson, J. L., Smith, B. O., Uhrín, D., Barlow, P. N., & Winder, S. J. (2002). Solution structure of the calponin CH domain and fitting to the 3D-helical reconstruction of F-actin:calponin. *Structure, Feb;10(2):249-58*. https://doi.org/10.1016/s0969-2126(02)00703-7.

Bronner, D. N., Abuaita, B. H., Chen, X., Fitzgerald, K. A., Nunez, G., He, Y., Yin, X. M., & O'Riordan, M. X. (2015). Endoplasmic Reticulum Stress Activates the Inflammasome via NLRP3- and Caspase-2-Driven Mitochondrial Damage. *Immunity*, *43*(3), 451–462.

Burger, D., Fickentscher, C., Moerloose, P., & Brandt, K. J. (2016). F-actin dampens NLRP3 inflammasome activity via Flightless-I and LRRFIP2. *Sci Rep*, *6*, 29834. doi: 10.1038/srep29834

Burgstaller, G., Kranewitter, W. J., & Gimona, M. (2002). The molecular basis for the autoregulation of calponin by isoform-specific C-terminal tail sequences. *J Cell Sci, 15;115(Pt 10):2021-9*. https://doi.org/10.1242/jcs.115.10.2021.

Busso, N., & So, A. (2012). Microcrystals as DAMPs and their role in joint inflammation. *Rheumatology (Oxford*, *51*(7), 1154–1160.

Butt, L., Unnersjo-Jess, D., Hohne, M., Edwards, A., Binz-Lotter, J., Reilly, D., Hahnfeldt, R., Ziegler, V., Fremter, K., Rinschen, M. M., Helmstadter, M., Ebert, L. K., Castrop, H., Hackl, M. J., Walz, G., Brinkkoetter, P. T., Liebau, M. C., Tory, K., Hoyer, P. F., … Benzing, T. (2020). A molecular mechanism explaining albuminuria in kidney disease. *Nat Metab*, *2*, 461-474,.

Cai, M. J., Liu, W., & Pei, X. Y. (2014). Juvenile hormone prevents 20-hydroxyecdysone-induced metamorphosis by regulating the phosphorylation of a newly identified broad protein. *J Biol Chem*, *289*(38), 26630–26641.

Caira, F. C., Stock, G., SR, TG, M., EC, H., J, B., & R.O. (2006). Human degenerative valve disease is associated with up-regulation of low-density lipoprotein receptor-related protein 5 receptor-mediated bone formation. *J Am Coll Cardiol*, *47*(8), 1707–1712.

Caligiuri, G., Nicoletti, A., Zhou, X., Tornberg, I., & Hansson, G. K. (1999). Effects of sex and age on atherosclerosis and autoimmunity in apoE-deficient mice. *Atherosclerosis*, *145*, 301–308.

Calle, Y., Burns, S., Thrasher, A. J., & Jones, G. E. (2006). The leukocyte podosome. *European Journal of Cell Biology*, *85*, 151-157,.

Camoretti-Mercado, B., Forsythe, S. M., & LeBeau, M. M. (1998). Expression and cytogenetic localization of the human SM22 gene (TAGLN. *Genomics*, *49*(3), 452–457.

Capobianco, A., Cottone, L., Monno, A., Manfredi, A. A., & Rovere-Querini, P. (2017). The peritoneum: Healing, immunity, and diseases. *J Pathol. Oct*, *243*(2), 137–147. https://doi.org/10.1002/path.4942.

Cardilo-Reis, L., Gruber, S., Schreier, S. M., Drechsler, M., Papac-Milicevic, N., Weber, C., Wagner, O., Stangl, H., Soehnlein, O., & Binder, C. J. (2012). Interleukin-13 protects from atherosclerosis and modulates plaque composition by skewing the macrophage phenotype. *EMBO Mol Med*, *4*, 1072–1086.

Castoldi, G., Gioia, C. R., Pieruzzi, F., Greef, W. M., Busca, G., Sperti, G., & Stella, A. (2001). Angiotensin II modulates calponin gene expression in rat vascular smooth muscle cells in vivo. *J Hypertens*, *19*, 2011–2018.

Castro, M. M., Cena, J., Cho, W. J., Walsh, M. P., & Schulz, R. (2012). Matrix metalloproteinase-2 proteolysis of calponin-1 contributes to vascular hypocontractility in endotoxemic rats. *Arterioscler Thromb Vasc Biol*, *32*, 662–668.

Cerecedo, D., Cisneros, B., Mondragon, R., Gonzalez, S., & Galvan, I. J. (2010). Actin filaments and microtubule dual-granule transport in human adhered platelets: The role of alpha-dystrobrevins. *Br J Haematol*, *149*, 124-136.

Cetin, O., Karaman, E., Boza, B., Cim, N., & Sahin, H. G. (2018). Maternal serum calponin 1 level as a biomarker for the short-term prediction of preterm birth in women with threatened preterm labor. *J Matern Fetal Neonatal Med*, *Jan;31(2):216–222.*

Chakrabarty, B., Lee, S., & Exintaris, B. (2019). Generation and Regulation of Spontaneous Contractions in the Prostate. *Adv Exp Med Biol*, *1124*, 195–215.

Chambers, J. E., Dalton, L. E., Clarke, H. J., Malzer, E., Dominicus, C. S., Patel, V., Moorhead, G., Ron, D., & Marciniak, S. J. (2015). Actin dynamics tune the integrated stress response by regulating eukaryotic initiation factor 2alpha dephosphorylation. *Elife*, *4*, e04872. doi: 10.7554/eLife.04872.

Chen, G. Y., & Nunez, G. (2010). Sterile inflammation: Sensing and reacting to damage. *Nat Rev Immunol*, *10*(12), 826–837.

Chen, J. H., Chen, W. L., Sider, K. L., Yip, C. Y., & Simmons, C. A. (2011). Beta-catenin mediates mechanically regulated, transforming growth factor-beta1-induced myofibroblast differentiation of aortic valve interstitial cells. *Arterioscler Thromb Vasc Biol*, *31*(3), 590–597.

Chen, J. H., & Simmons, C. A. (2011). Cell-matrix interactions in the pathobiology of calcific aortic valve disease: Critical roles for matricellular, matricrine, and matrix mechanics cues. *Circ Res*, *108*(12), 1510–1524.

Chen, J., & Zolkiewska, A. (2011). Force-induced unfolding simulations of the human Notch1 negative regulatory region: Possible roles of the heterodimerization domain in mechanosensing. *PLoS One*, *6*(7), e22837. doi: 10.1371/journal.pone.0022837.

Chen, P., Cescon, M., & Bonaldo, P. (2014). Autophagy-mediated regulation of macrophages and its applications for cancer. *Autophagy*, *10*(2), 192–200.

Cheng, Y., Feng, Y., Jansson, L., Sato, Y., Deguchi, M., Kawamura, K., & Hsueh, A. J. (2015). Actin polymerization-enhancing drugs promote ovarian follicle growth mediated by the Hippo signaling effector YAP. *FASEB Journal: Official Publication of the Federation of American Societies for Experimental Biology*, *29*(6), 2423–2430. https://doi.org/10.1096/fj.14-267856.

Chiba, T., Ikeda, M., Umegaki, K., & Tomita, T. (2011). Estrogen-dependent activation of neutral cholesterol ester hydrolase underlying gender difference of atherogenesis in apoE-/- mice. *Atherosclerosis*, *219*, 545–551.

Chicurel, M. E., CS, C., & Ingber, D. E. (1998). Cellular control lies in the balance of forces. *Curr Opin Cell Biol*, *10*, 232–239.

Childs, T. J., Watson, M. H., Novy, R. E., Lin, J. J. C., & Mak, A. S. (1992). Calponin and tropomyosin interactions. *Biochimica et Biophysica Acta (BBA)/Protein Structure and Molecular*, *1121*(1–2), 41–46. https://doi.org/10.1016/0167-4838(92)90334-A.

Chinetti-Gbaguidi, G., Colin, S., & Staels, B. (2015). Macrophage subsets in atherosclerosis. *Nat Rev Cardiol*, *12*, 10–17.

Chiu, J. J., & Chien, S. (2011). Effects of disturbed flow on vascular endothelium: Pathophysiological basis and clinical perspectives. *Physiol Rev*, *91*(1), 327–387.

Chon, S. J., Umair, Z., & Yoon, M.-S. (2021). Premature Ovarian Insufficiency: Past, Present, and Future. *Frontiers in Cell and Developmental Biology*, *9*, 672890. https://doi.org/10.3389/fcell.2021.672890.

Chong, A. S., Parish, C. R., & Coombe, D. R. (1987). Evidence that the cytoskeleton plays a key role in cell adhesion. *Immunol Cell Biol*, *65*(Pt 1), 85–95.

Choquet, D., Felsenfeld, D. P., & Sheetz, M. P. (1997). Extracellular matrix rigidity causes strengthening of integrin-cytoskeleton linkages. *Cell*, *88*, 39–48.

Cirka, H. A., Kural, M. H., & Billiar, K. L. (2015). Mechanoregulation of aortic valvular interstitial cell life and death. *J Long Term Eff Med Implants*, *25*(1–2), 3–16.

Clausen, B. E., Burkhardt, C., Reith, W., Renkawitz, R., & Forster, I. (1999). Conditional gene targeting in macrophages and granulocytes using LysMcre mice. *Transgenic Res*, *8*, 265-277.

Coelho Neto, M. A., Ludwin, A., Borrell, A., Benacerraf, B., Dewailly, D., da Silva Costa, F., Condous, G., Alcazar, J. L., Jokubkiene, L., Guerriero, S., Van den Bosch, T., & Martins, W. P. (2018). Counting ovarian antral follicles by ultrasound: A practical guide. *Ultrasound in Obstetrics & Gynecology: The Official Journal of the International Society of Ultrasound in Obstetrics and Gynecology*, *51*(1), 10–20. https://doi.org/10.1002/uog.18945.

Coleman, R., Hayek, T., Keidar, S., & Aviram, M. (2006). A mouse model for human atherosclerosis: Long-term histopathological study of lesion development in the aortic arch of apolipoprotein E-deficient (E0) mice. *Acta Histochem*, *108*, 415–424.

Colgren, J., & Nichols, S. A. (2022). MRTF specifies a muscle-like contractile module in Porifera. *Nat Commun*, *13*(1), 4134. doi: 10.1038/s41467-022-31756-9.

Craici, I. M., Wagner, S. J., Weissgerber, T. L., Grande, J. P., & Garovic, V. D. (2014). Advances in the pathophysiology of pre-eclampsia and related podocyte injury. *Kidney Int*, *86*, 275-285.

Czurylo, E. A., Kulikova, N., & Dbrowska, R. (1997). Does calponin interact with caldesmon? *J Biol Chem*, *272*, 32067–32070.

Daimon, E., Shibukawa, Y., & Wada, Y. (2013). Calponin 3 regulates stress fiber formation in dermal fibroblasts during wound healing. *Archives of Dermatological Research*, *305*(7), 571–584. https://doi.org/10.1007/s00403-013-1343-8.

Dandekar, A., Mendez, R., & Zhang, K. (2015). Cross talk between ER stress, oxidative stress, and inflammation in health and disease. *Methods Mol Biol*, *1292*, 205–214.

Danninger, C., & Gimona, M. (2000). Live dynamics of GFP-calponin: Isoform-specific modulation of the actin cytoskeleton and autoregulation by C-terminal sequences. *J Cell Sci*, *21*, 3725–3736.

Das, R., Ganapathy, S., Settle, M., & Plow, E. F. (2014). Plasminogen promotes macrophage phagocytosis in mice. *Blood*, *124*(5), 679–688.

Daugherty, A. (2002). Mouse models of atherosclerosis. *Am J Med Sci*, *323*, 3–10.

Davies, L. C., Jenkins, S. J., Allen, J. E., Taylor, P. R., & PF, G. M. A. (2013). Tissue-resident macrophages. *Nat Immunol*, *14*(10), 986–995.

Davies, P. F., & Guerraty, M. A. (2011). Getting physical with the aortic valve. *Arteriosclerosis, Thrombosis, and Vascular Biology*, *31*(3), 474–475. https://doi.org/10.1161/ATVBAHA.110.220962.

Davis, M. J., Wu, X., Nurkiewicz, T. R., Kawasaki, J., Davis, G. E., Hill, M. A., & Meininger, G. A. (2001). Integrins and mechanotransduction of the vascular myogenic response. *American Journal of Physiology. Heart and Circulatory Physiology*, *280*(4), 1427–1433.

de Cavanagh, E. M., Ferder, M., Inserra, F., & Ferder, L. (2009). Angiotensin II, mitochondria, cytoskeletal, and extracellular matrix connections: An integrating viewpoint. *American Journal of Physiology. Heart and Circulatory Physiology*, *296*(3), H550-558. https://doi.org/10.1152/ajpheart.01176.2008.

De, R., Zemel, A., & Safran, S. A. (2010). Theoretical concepts and models of cellular mechanosensing. *Methods in Cell Biology*, *98*, 143–175. https://doi.org/10.1016/S0091-679X(10)98007-2.

Debald, M., Jin, J. P., Linke, A., Walgenbach, K. J., Rauch, P., Zellmer, A., Fimmers, R., Kuhn, W., Hartmann, G., & Walgenbach-Brünagel, G. (2014). Calponin-h2: A potential serum marker for the early detection of human breast cancer? *Tumor Biology*, *35*(11), 11121–11127. https://doi.org/10.1007/s13277-014-2419-6.

Di Giuseppe, M., Gambelli, F., Hoyle, G. W., Lungarella, G., Studer, S. M., Richards, T., Yousem, S., McCurry, K., Dauber, J., Kaminski, N., Leikauf, G., & Ortiz, L. A. (2009). Systemic inhibition of NF-kappaB activation protects from silicosis. *PloS One*, *4*(5), e5689. https://doi.org/10.1371/journal.pone.0005689.

Dickhout, J. G., Basseri, S., & Austin, R. C. (2008). Macrophage function and its impact on atherosclerotic lesion composition, progression, and stability: The good, the bad, and the ugly. *Arterioscler Thromb Vasc Biol*, *28*, 1413–1415.

Ding, R., Li, G., & Yao, Y. (2022). Transgelin-2 interacts with CD44 to regulate Notch1 signaling pathway and participates in colorectal cancer proliferation and migration. *J Physiol Biochem*, *78*(1), 99–108.

Dobrokhotov, O., Samsonov, M., Sokabe, M., & Hirata, H. (2018) Mechanoregulation and pathology of YAP/TAZ via Hippo and non-Hippo mechanisms. *Clin Transl Med*. 7(1), 23. doi: 10.1186/s40169-018-0202-9.

Dobrzhanskaya, A. V., Vyatchin, I. G., Lazarev, S. S., Matusovsky, O. S., & Shelud'ko, N. S. (2013). Molluscan smooth catch muscle contains calponin but not caldesmon. *J Muscle Res Cell Motil*, *34*(1), 23–33.

Dose, J., Huebbe, P., Nebel, A., & Rimbach, G. (2016). APOE genotype and stress response—A mini review. *Lipids Health Dis*, *15*, 121. doi: 10.1186/s12944-016-0288-2.

Dostert, C., Petrilli, V., Bruggen, R., Steele, C., Mossman, B. T., & Tschopp, J. (2008). Innate immune activation through Nalp3 inflammasome sensing of asbestos and silica. *Science, 320*(5876), 674–677.

Draeger, A., Gimona, M., Stuckert, A., Celis, J. E., & Small, J. V. (1991). Calponin. Developmental isoforms and a low molecular weight variant. *FEBS Letters, 291*(1), 24–28. https://doi.org/10.1016/0014-5793(91)81095-p.

Duan, B., Yin, Z., Hockaday Kang, L., Magin, R. L., & Butcher, J. T. (2016). Active tissue stiffness modulation controls valve interstitial cell phenotype and osteogenic potential in 3D culture. *Acta Biomater, 36*, 42–54.

Duband, J. L., Gimona, M., Scatena, M., Sartore, S., & Small, J. V. (1993). Calponin and SM 22 as differentiation markers of smooth muscle: Spatiotemporal distribution during avian embryonic development. *Differentiation, 55*(1), 1–11.

Dulin, N. O., Orlov, S. N., Kitchen, C. M., Voyno-Yasenetskaya, T. A., & Miano, J. M. (2001). G-protein-coupled-receptor activation of the smooth muscle calponin gene. *Biochem J, 357*, 587–592.

Duncan, M., Cummings, L., & Chada, K. (1993). Germ cell deficient (gcd) mouse as a model of premature ovarian failure. *Biology of Reproduction, 49*(2), 221–227. https://doi.org/10.1095/biolreprod49.2.221.

Dvorakova, M., Jerabkova, J., & Prochazkova, I. (2016). Transgelin is upregulated in stromal cells of lymph node positive breast cancer. *J Proteomics, 132*, 103–111.

Dweck, M. R., Boon, N. A., & Newby, D. E. (2012). Calcific aortic stenosis: A disease of the valve and the myocardium. *J Am Coll Cardiol, 60*(19), 1854–1863.

Dweck, M. R., Khaw, H. J., Sng, G. K., Luo, E. L., Baird, A., & Williams, M. C. (2013). Aortic stenosis, atherosclerosis, and skeletal bone: Is there a common link with calcification and inflammation? *Eur Heart J, 34*(21), 1567–1574.

Dykes, A. C., & Wright, G. L. (2007). Down-regulation of calponin destabilizes actin cytoskeletal structure in A7r5 cells. *Can J Physiol Pharmacol, 85*, 225–232.

Eckes, B., & Krieg, T. (2004). Regulation of connective tissue homeostasis in the skin by mechanical forces. *Clinical and Experimental Rheumatology, 22*(3 Suppl 33), S73-76.

EL-Mezgueldi, M., & Marston, S. B. (1996). The effects of smooth muscle calponin on the strong and weak myosin binding sites of F-actin. *The Journal of Biological Chemistry, 271*(45), 28161–28167. https://doi.org/10.1074/jbc.271.45.28161.

Elsafadi, M., Manikandan, M., & Almalki, S. (2020). Transgelin is a poor prognostic factor associated with advanced colorectal cancer (CRC) stage promoting tumor growth and migration in a TGFbeta-dependent manner. *Cell Death Dis, 11*(5), 341. doi: 10.1038/s41419-020-2529-6.

Elsafadi, M., Manikandan, M., & Dawud, R. A. (2016). Transgelin is a TGFbeta-inducible gene that regulates osteoblastic and adipogenic differentiation of human skeletal stem cells through actin cytoskeleston organization. *Cell Death Dis, 7*(8), e2321. doi: 10.1038/cddis.2016.196.

Endlich, N., & Endlich, K. (2006). Stretch, tension and adhesion—Adaptive mechanisms of the actin cytoskeleton in podocytes. *Eur J Cell Biol, 85*, 229-234.

Eswarappa, S. M., Pareek, V., & Chakravortty, D. (2008). Role of actin cytoskeleton in LPS-induced NF-kappaB activation and nitric oxide production in murine macrophages. *Innate Immun, 14*(5), 309–318.

Facemire, C., Brozovich, F. V., & Jin, J. P. (2000). The maximal velocity of vascular smooth muscle shortening is independent of the expression of calponin. *Journal of Muscle Research and Cell Motility*, *21*(4), 367–373. https://doi.org/10.1023/A:1005680614296.

Fan, L., Jaquet, V., & Dodd, P. R. (2001). Molecular cloning and characterization of hNP22: A gene up-regulated in human alcoholic brain. *J Neurochem*, *76*, 1275–1281.

Farzaneh-Far, A., Proudfoot, D., Shanahan, C., & Weissberg, P. L. (2001). Vascular and valvar calcification: Recent advances. *Heart*, *85*(1), 13–17.

Fattoum, A., Roustan, C., Smyczynski, C., Der Terrossian, E., Kassab, R. (2003) Mapping the microtubule binding regions of calponin. *Biochemistry* 42(5), 1274-82.

Fedorchak, G. R., Kaminski, A., & Lammerding, J. (2014). Cellular mechanosensing: Getting to the nucleus of it all. *Progress in Biophysics and Molecular Biology*, *115*(2–3), 76–92. https://doi.org/10.1016/j.pbiomolbio.2014.06.009.

Feng, H. Z., Wang, H., Takahashi, K., & Jin, J. P. (2019). Double deletion of calponin 1 and calponin 2 in mice decreases systemic blood pressure with blunted length-tension response of aortic smooth muscle. *Journal of Molecular and Cellular Cardiology*, *129*, 49–57. https://doi.org/10.1016/j.yjmcc.2019.01.026.

Ferhat, L., Charton, G., Represa, A., Ben-Ari, Y., Terrossian, E., & Khrestchatisky, M. (1996). Acidic calponin cloned from neural cells is differentially expressed during rat brain development. *European Journal of Neuroscience*, *8*(7), 1501–1509. https://doi.org/10.1111/j.1460-9568.1996.tb01612.x.

Ferhat, L., Esclapez, M., Represa, A., Fattoum, A., Shirao, T., & Ben-Ari, Y. (2003). Increased levels of acidic calponin during dendritic spine plasticity after pilocarpine-induced seizures. *Hippocampus*, *13*, 845–858.

Ferjani, I., Fattoum, A., Maciver, S. K., Benistant, C., Chahinian, A., Manai, M., Benyamin, Y., & Roustan, C. (2006). A direct interaction with calponin inhibits the actin-nucleating activity of gelsolin. *Biochem J*, *396*, 461-468.

Ferjani, I., Fattoum, A., Manai, M., Benyamin, Y., Roustan, C., & Maciver, S. K. (2010). Two distinct regions of calponin share common binding sites on actin resulting in different modes of calponin-actin interaction. *Biochim Biophys Acta*, 1760–1767.

Ferrari, A. U., Radaelli, A., Mori, T., Mircoli, L., Perlini, S., Meregalli, P., Fedele, L., & Mancia, G. (2001). Nitric oxide-dependent vasodilation and the regulation of arterial blood pressure. *J Cardiovasc Pharmacol*, *38 Suppl 2*, 19–22.

Fidler, I. J. (1990). Critical Factors in the Biology of Human Cancer Metastasis: Twenty-eighth G. *Cancer Res*, *50*, 6130–6138.

Fieren, M. W. (2012). The local inflammatory responses to infection of the peritoneal cavity in humans: Their regulation by cytokines, macrophages, and other leukocytes. *Mediators Inflamm*, *2012*, 976241. doi: 10.1155/2012/976241.

Flemming, A., Huang, Q. Q., Jin, J. P., Jumaa, H., & Herzog, S. (2015). A conditional knockout mouse model reveals that Calponin-3 is dispensable for early B cell development. *PLoS ONE*, *10*(6), 1–16. https://doi.org/10.1371/journal.pone.0128385.

Folkow, B. (1989). Myogenic mechanisms in the control of systemic resistance. *Introduction and Historical Background, J Hypertens Suppl*, 7(4), 1–4.

Fontao, L., Geerts, D., Kuikman, I., Koster, J., Kramer, D., & Sonnenberg, A. (2001). The interaction of plectin with actin: Evidence for cross-linking of actin filaments by dimerization of the actin-binding domain of plectin. *J Cell Sci*, *114*, 2065–2076.

Franchi, L., Eigenbrod, T., & Nunez, G. (2009). Cutting edge: TNF-alpha mediates sensitization to ATP and silica via the NLRP3 inflammasome in the absence of microbial stimulation. *J Immunol*, *183*(2), 792–796.

Fraser, E. D., & Walsh, M. P. (1995). Dephosphorylation of calponin by type 2B protein phosphatase. *Biochemistry*, *34*(28), 9151–9158. https://doi.org/10.1021/bi00028a026.

Freigang, S., Ampenberger, F., Weiss, A., Kanneganti, T. D., Iwakura, Y., Hersberger, M., & Kopf, M. (2013). Fatty acid-induced mitochondrial uncoupling elicits inflammasome-independent IL-1alpha and sterile vascular inflammation in atherosclerosis. *Nat Immunol*, *14*, 1045–1053.

Fritz, M., & Rinaldi, G. (2007). Influence of nitric oxide-mediated vasodilation on the blood pressure measured with the tail-cuff method in the rat. *J Biomed Sci*, *14*(6), 757–765.

Fu, Q., Liu, P. C., Wang, J. X., Song, Q. S., & Zhao, X. F. (2009). Proteomic identification of differentially expressed and phosphorylated proteins in epidermis involved in larval-pupal metamorphosis of Helicoverpa armigera. *BMC Genomics*, *10*, 600. doi: 10.1186/1471-2164-10-600.

Fu, Y., Liu, H. W., Forsythe, S. M., Kogut, P., McConville, J. F., Halayko, A. J., Camoretti-Mercado, B., & Solway, J. (2000). Mutagenesis analysis of human SM22: Characterization of actin binding. *Journal of Applied Physiology (Bethesda, Md.: 1985)*, *89*(5), 1985–1990. https://doi.org/10.1152/jappl.2000.89.5.1985.

Fujii, T., & Koizumi, Y. (1999). Identification of the binding region of basic calponin on alpha and beta tubulins. *Journal of Biochemistry*, *125*(5), 869–875. https://doi.org/10.1093/oxfordjournals.jbchem.a022362.

Fujii, T., Oomatsuzawa, A., Kuzumaki, N., & Kondo, Y. (1994). Calcium-dependent regulation of smooth muscle calponin by S100. *J Biochem*, *116*, 121–127.

Fujishige, A., Takahashi, K., & Tsuchiya, T. (2002). Altered mechanical properties in smooth muscle of mice with a mutated calponin locus. *Zoolog Sci*, *19*(2), 167–174.

Fukui, Y., Masuda, H., Takagi, M., Takahashi, K., & Kiyokane, K. (1997). The presence of h2-calponin in human keratinocyte. *J Dermatol Sci*, *14*, 29–36.

Galkin, V. E., Orlova, A., Fattoum, A., Walsh, M. P., & Egelman, E. H. (2006). The CH-domain of calponin does not determine the modes of calponin binding to F-actin. *J Mol Biol*, *359*(2), 478–485.

Galli, S. J., Borregaard, N., & Wynn, T. A. (2011). Phenotypic and functional plasticity of cells of innate immunity: Macrophages, mast cells and neutrophils. *Nature Immunology*, *12*, 1035-1044.

Gao, G., Chen, L., Dong, B., Gu, H., Dong, H., Pan, Y., Gao, Y., & Chen, X. (2009). RhoA effector mDia1 is required for PI 3-kinase-dependent actin remodeling and spreading by thrombin in platelets. *Biochemical and Biophysical Research Communications*, *385*, 439-444.

Gao, J., Hwang, J. M., & Jin, J. P. (1996). Complete nucleotide sequence, structural organization, and an alternatively spliced exon of mouse h1-calponin gene. *Biochemical and Biophysical Research Communications*, *218*(1), 292–297. https://doi.org/10.1006/bbrc.1996.0051.

Gaytán, F., Morales, C., Manfredi-Lozano, M., & Tena-Sempere, M. (2014). Generation of multi-oocyte follicles in the peripubertal rat ovary: Link to the invasive capacity of granulosa cells? *Fertility and Sterility*, *101*(5), 1467–1476. https://doi.org/10.1016/j.fertnstert.2014.01.037.

Gerdes, M. J., Larsen, M., Dang, T. D., Ressler, S. J., JA, T., & Rowley, D. R. (2004). Regulation of rat prostate stromal cell myodifferentiation by androgen and TGF-beta1. *Prostate*, *58*, 299–307.

Gerthoffer, W. T., & Pohl, J. (1994). Caldesmon and calponin phosphorylation in regulation of smooth muscle contraction. *Can J Physiol Pharmacol*, *72*, 1410–1414.

Gimona, M., Djinovic-Carugo, K., Kranewitter, W. J., & Winder, S. J. (2002). Functional plasticity of CH domains. *FEBS Letters*, *513*(1), 98–106. https://doi.org/10.1016/S0014-5793(01)03240-9.

Gimona, M., Kaverina, I., & Resch, G. P. (2003). Calponin repeats regulate actin filament stability and formation of podosomes in smooth muscle cells. *Mol Biol Cell*, *14*, 2482–2491.

Gimona, M., Sparrow, M. P., Strasser, P., Herzog, M., & Small, J. V. (1992). Calponin and SM 22 isoforms in avian and mammalian smooth muscle: Absence of phosphorylation in vivo. *European Journal of Biochemistry*, *205*(3), 1067–1075. https://doi.org/10.1111/j.1432-1033.1992.tb16875.x.

Goetinck, S., & Waterston, R. H. (1994). The Caenorhabditis elegans muscle-affecting gene unc-87 encodes a novel thin filament-associated protein. *J Cell Biol*, *127*(1), 79–93.

Gonzalez, M. M., Yoshizaki, L., Wolfenstein-Todel, C., & Fink, N. E. (2012). Isolation of galectin-1 from human platelets: Its interaction with actin. *Protein J*, *31*, 8-14.

Goodman, A., Goode, B. L., Matsudaira, P., & Fink, G. R. (2003). The Saccharomyces cerevisiae calponin/transgelin homolog Scp1 functions with fimbrin to regulate stability and organization of the actin cytoskeleton. *Mol Biol Cell*, *14*(7), 2617–2629.

Gordon, S., Hamann, J., Lin, H. H., & Stacey, M. (2011). F4/80 and the related adhesion-GPCRs. *Eur J Immunol*, *41*(9), 2472–2476.

Gordon, S., & Pluddemann, A. (2017). Tissue macrophages: Heterogeneity and functions. *BMC Biol*, *15*(1), 53. doi: 10.1186/s12915-017-0392-4.

Gould, S. T., Srigunapalan, S., Simmons, C. A., & Anseth, K. S. (2013). Hemodynamic and cellular response feedback in calcific aortic valve disease. *Circ Res*, *113*(2), 186–197.

Gozal, E., Ortiz, L. A., Zou, X., Burow, M. E., Lasky, J. A., & Friedman, M. (2002). Silica-induced apoptosis in murine macrophage: Involvement of tumor necrosis factor-alpha and nuclear factor-kappaB activation. *Am J Respir Cell Mol Biol*, *27*(1), 91–98.

Graceffa, P., Adam, L. P., & Morgan, K. G. (1996). Strong interaction between caldesmon and calponin. *Journal of Biological Chemistry*, *271*(48), 30336–30339. https://doi.org/10.1074/jbc.271.48.30336.

Greenlee, R. T., Murray, T., S, B., & Wingo, P. A. (2000). Cancer statistics. *CA Cancer J Clin*, *50*, 7–33.

Grigoriev, V. G., Thweatt, R., & Moerman, E. J. (1996). Expression of senescence-induced protein WS3-10 in vivo and in vitro. *Exp Gerontol*, *31*, 145–157.

Grootjans, J., Kaser, A., Kaufman, R. J., & Blumberg, R. S. (2016). The unfolded protein response in immunity and inflammation. *Nat Rev Immunol*, *16*(8), 469–484.

Guest, T. M., Vlastos, G., Alameddine, F. M. F., & Taylor, W. R. (2006). Mechanoregulation of monocyte chemoattractant protein-1 expression in rat vascular smooth muscle cells. *Antioxidants & Redox Signaling*, *8*(9–10), 1461–1471. https://doi.org/10.1089/ars.2006.8.1461.

Gusev, N. B. (2001). Some properties of caldesmon and calponin and the participation of these proteins in regulation of smooth muscle contraction and cytoskeleton formation. *Biochemistry (Mosc*, *66*(10), 1112–1121.

Hackbarth, H., & Harrison, D. E. (1982). Changes with age in renal function and morphology in C57BL/6, CBA/HT6, and B6CBAF1 mice. *J Gerontol*, *37*, 540-547.

Haeberle, J. (1994). Calponin decreases the rate of cross-bridge cycling and increases maximum force production by smooth muscle myosin in an in vitro motility assay. *J Biol Chem*, *269*(17), 12424–12431.

Haidl, I. D., & Jefferies, W. A. (1996). The macrophage cell surface glycoprotein F4/80 is a highly glycosylated proteoglycan. *Eur J Immunol*, *26*(5), 1139–1146.

Hallett, J. M., Leitch, A. E., Riley, N. A., Duffin, R., Haslett, C., & Rossi, A. G. (2008). Novel pharmacological strategies for driving inflammatory cell apoptosis and enhancing the resolution of inflammation. *Trends Pharmacol Sci*, *29*, 250-257.

Hamers, A. A., Vos, M., Rassam, F., Marinkovic, G., Kurakula, K., Gorp, P. J., Winther, M. P., Gijbels, M. J., Waard, V., & Vries, C. J. (2012). Bone marrow-specific deficiency of nuclear receptor Nur77 enhances atherosclerosis. *Circ Res*, *110*, 428–438.

Hamill, O. P., & Martinac, B. (2001). Molecular basis of mechanotransduction in living cells. *Physiological Reviews*, *81*(2), 685–740. https://doi.org/10.1152/physrev.2001.81.2.685.

Han, E. K. H., Guadagno, T. M., SL, D., & Assoian, R. K. (1993). A cell cycle and mutational analysis of anchorage-dependent growth: Cell adhesion and TGF-beta1 control G1/S transit specificity. *J Cell Biol*, *122*, 461–471.

Han, Y. H., Ryu, K. B., Medina Jimenez, B. I., Kim, J., Lee, H. Y., & Cho, S. J. (2020). Muscular Development in Urechis unicinctus (Echiura, Annelida. *Int J Mol Sci*, *21*(7), 2306. doi: 10.3390/ijms21072306.

Hanna, R. N., Shaked, I., Hubbeling, H. G., Punt, J. A., Wu, R., Herrley, E., Zaugg, C., Pei, H., Geissmann, F., Ley, K., & Hedrick, C. C. (2012). NR4A1 (Nur77) deletion polarizes macrophages toward an inflammatory phenotype and increases atherosclerosis. *Circ Res*, *110*, 416–427.

Harre, U., Georgess, D., Bang, H., Bozec, A., Axmann, R., Ossipova, E., Jakobsson, P. J., Baum, W., Nimmerjahn, F., Szarka, E., Sarmay, G., Krumbholz, G., Neumann, E., Toes, R., Scherer, H. U., Catrina, A. I., Klareskog, L., Jurdic, P., & Schett, G. (2012). Induction of osteoclastogenesis and bone loss by human autoantibodies against citrullinated vimentin. *The Journal of Clinical Investigation*, *122*, 1791-1802.

Hasegawa, M., Moritani, S., Murakumo, Y., Sato, T., Hagiwara, S., Suzuki, C., Mii, S., Jijiwa, M., Enomoto, A., Asai, N., Ichihara, S., & Takahashi, M. (2008). CD109 expression in basal-like breast carcinoma. *Pathol Int*, *58*, 288–294.

Hellstrand, P., & Albinsson, S. (2005). Stretch-dependent growth and differentiation in vascular smooth muscle: Role of the actin cytoskeleton. *Canadian Journal of Physiology and Pharmacology*, *83*(10), 869–875. https://doi.org/10.1139/y05-061.

Hines, P. C., Gao, X., White, J. C., D'Agostino, A., & Jin, J. P. (2014). A novel role of h2-calponin in regulating whole blood thrombosis and platelet adhesion during physiologic flow. *Physiological Reports*, *2*(12), 1–11. https://doi.org/10.14814/phy2.12228.

Hinz, B. (2016). Myofibroblasts. *Experimental Eye Research*, *142*, 56–70. https://doi.org/10.1016/j.exer.2015.07.009.

Hirsch, S., Austyn, J. M., & Gordon, S. (1981). Expression of the macrophage-specific antigen F4/80 during differentiation of mouse bone marrow cells in culture. *J Exp Med*, *154*(3), 713–725.

Hjortnaes, J., Goettsch, C., Hutcheson, J. D., Camci-Unal, G., Lax, L., & Scherer, K. (2016). Simulation of early calcific aortic valve disease in a 3D platform: A role for myofibroblast differentiation. *J Mol Cell Cardiol*, *94*, 13–20.

Hodgin, J. B., Bitzer, M., Wickman, L., Afshinnia, F., Wang, S. Q., O'Connor, C., Yang, Y., Meadowbrooke, C., Chowdhury, M., Kikuchi, M., Wiggins, J. E., & Wiggins, R. C. (2015). Glomerular Aging and Focal Global Glomerulosclerosis: A Podometric Perspective. *J Am Soc Nephrol*, *26*, 3162-3178.

Hodgkinson, J. L., el-Mezgueldi, M., Craig, R., Vibert, P., Marston, S. B., & Lehman, W. (1997). 3-D image reconstruction of reconstituted smooth muscle thin filaments containing calponin: Visualization of interactions between F-actin and calponin. *J Mol Biol*, *17;273(1):150-9*. https://doi.org/10.1006/jmbi.1997.1307.

Holm, P. W., Slart, R. H., Zeebregts, C. J., Hillebrands, J. L., & Tio, R. A. (2009). Atherosclerotic plaque development and instability: A dual role for VEGF. *Ann Med*, *41*, 257–264.

Horiuchi, A., Nikaido, T., Ito, K., Zhai, Y., Orii, A., Taniguchi, S., Toki, T., & Fujii, S. (1998). Reduced expression of calponin h1 in leiomyosarcoma of the uterus. *Lab Invest*, *78*, 839–846.

Horiuchi, A., Nikaido, T., Taniguchi, S., & Fujii, S. (1999). Possible role of calponin h1 as a tumor suppressor in human uterine leiomyosarcoma. *J Natl Cancer Inst*, *91*, 790–796.

Hornung, V., Bauernfeind, F., Halle, A., Samstad, E. O., Kono, H., Rock, K. L., Fitzgerald, K. A., & Latz, E. (2008). Silica crystals and aluminum salts activate the NALP3 inflammasome through phagosomal destabilization. *Nat Immunol*, *9*(8), 847–856.

Hossain, M. M., Crish, J. F., Eckert, R. L., Lin, J. J.-C., & Jin, J.-P. (2005). H2-Calponin is regulated by mechanical tension and modifies the function of actin cytoskeleton. *The Journal of Biological Chemistry*, *280*(51), 42442–42453. https://doi.org/10.1074/jbc.M509952200.

Hossain, M. M., Hwang, D. Y., Huang, Q. Q., Sasaki, Y., & Jin, J. P. (2003). Developmentally regulated expression of calponin isoforms and the effect of h2-calponin on cell proliferation. *American Journal of Physiology - Cell Physiology*, *284*(1 53-1), 156–167. https://doi.org/10.1152/ajpcell.00233.2002.

Hossain, M. M., Smith, P. G., Wu, K., & Jin, J.-P. (2006). Cytoskeletal tension regulates both expression and degradation of h2-calponin in lung alveolar cells. *Biochemistry*, *45*(51), 15670–15683. https://doi.org/10.1021/bi061718f.

Hossain, M. M., Zhao, G., Woo, M. S., Wang, J. H. C., & Jin, J. P. (2016). Deletion of Calponin 2 in Mouse Fibroblasts Increases Myosin II-Dependent Cell Traction Force. *Biochemistry*, *55*(43), 6046–6055. https://doi.org/10.1021/acs.biochem.6b00856.

Hoyer-Hansen, M., & Jaattela, M. (2007). Connecting endoplasmic reticulum stress to autophagy by unfolded protein response and calcium. *Cell Death Differ*, *14*(9), 1576–1582.

Hsieh, T. B., Feng, H. Z., & Jin, J. P. (2022). Deletion of Calponin 2 Reduces the Formation of Postoperative Peritoneal Adhesions. *Journal of Investigative Surgery*, *35*(3), 517–524. https://doi.org/10.1080/08941939.2021.1880672.

Hsieh, T. B., & Jin, J. P. (2022). Loss of Calponin 2 causes age-progressive proteinuria in mice. *Physiological Reports*, *10*(18), 1–16. https://doi.org/10.14814/phy2.15370.

Hsieh, T.-B., & Jin, J.-P. (2023). Evolution and function of calponin and transgelin. *Frontiers in Cell and Developmental Biology*, *11*(June), 1–18. https://doi.org/10.3389/fcell.2023.1206147.

Hsieh, T.-B., & Jin, J.-P. (2024). Loss of Calponin 2 causes premature ovarian insufficiency in mice. *Journal of Ovarian Research*. https://doi.org/10.1186/s13048-024-01346-y.

Huang, Q. Q., Hossain, M. M., Sun, W., Xing, L., Pope, R. M., & Jin, J. P. (2016). Deletion of calponin 2 in macrophages attenuates the severity of inflammatory arthritis in mice. *American Journal of Physiology - Cell Physiology*, *311*(4), 673–685. https://doi.org/10.1152/ajpcell.00331.2015.

Huang, Q. Q., Hossain, M. M., Wu, K., Parai, K., Pope, R. M., & Jin, J. P. (2008). Role of H2-calponin in regulating macrophage motility and phagocytosis. *Journal of Biological Chemistry*, *283*(38), 25887–25899. https://doi.org/10.1074/jbc.M801163200.

Huang, Q. Q., Perlman, H., Huang, Z., Birkett, R., Kan, L., Agrawal, H., Misharin, A., Gurbuxani, S., Crispino, J. D., & Pope, R. M. (2010). FLIP: a novel regulator of macrophage differentiation and granulocyte homeostasis. *Blood*, *116*, 4968-4977.

Huber, P. A. (1997). Caldesmon. *The International Journal of Biochemistry & Cell Biology*, *29*(8–9), 1047–1051. https://doi.org/10.1016/s1357-2725(97)00004-6.

Hutcheson, J. D., Aikawa, E., & Merryman, W. D. (2014). Potential drug targets for calcific aortic valve disease. *Nat Rev Cardiol*, *11*(4), 218–231.

Ichikawa, K., Ito, M., Okubo, S., Konishi, T., Nakano, T., Mino, T., Nakamura, F., Naka, M., & Tanaka, T. (1993). Calponin phosphatase from smooth muscle: A possible role of type 1 protein phosphatase in smooth muscle relaxation. *Biochem Biophys Res Commun*, *193*, 827–833.

Ingber, D. (1999). How cells (might) sense microgravity. *FASEB J*, *13*, 3–15. https://doi.org/10.1096/fasebj.13.9001.s3.

Inoue, M., & Shinohara, M. L. (2014). Clustering of pattern recognition receptors for fungal detection. *PLoS Pathog*, *10*(2):e1003873. doi: 10.1371/journal.ppat.1003873.

Ishiwata-Kimata, Y., Yamamoto, Y. H., Takizawa, K., Kohno, K., & Kimata, Y. (2013). F-actin and a type-II myosin are required for efficient clustering of the ER stress sensor Ire1. *Cell Struct Funct*, *38*(2), 135–143.

Islam, A. H., Ehara, T., Kato, H., Hayama, M., & Nishizawa, O. (2004). Loss of calponin h1 in renal angiomyolipoma correlates with aggressive clinical behavior. *Urology*, *64*, 468–473.

Iung, B., Baron, G., Butchart, E. G., Delahaye, F., Gohlke-Bärwolf, C., Levang, O. W., Tornos, P., Vanoverschelde, J.-L., Vermeer, F., Boersma, E., Ravaud, P., & Vahanian, A. (2003). A prospective survey of patients with valvular heart disease in Europe: The Euro Heart Survey on Valvular Heart Disease. *European Heart Journal*, *24*(13), 1231–1243. https://doi.org/10.1016/s0195-668x(03)00201-x.

Iung, B., & Vahanian, A. (2011). Epidemiology of valvular heart disease in the adult. *Nat Rev Cardiol*, *8*(3), 162–172.

Ix, J. H., Shlipak, M. G., Katz, R., Budoff, M. J., Shavelle, D. M., & Probstfield, J. L. (2007). Kidney function and aortic valve and mitral annular calcification in the Multi-Ethnic Study of Atherosclerosis (MESA. *Am J Kidney Dis*, *50*(3), 412–420.

Jablonski, K. A., Amici, S. A., Webb, L. M., Ruiz-Rosado Jde, D., Popovich, P. G., Partida-Sanchez, S., & Guerau-de-Arellano, M. (2015). Novel Markers to Delineate Murine M1 and M2 Macrophages. *PLoS One*, *10*(12), e0145342. doi: 10.1371/journal.pone.0145342.

Jagarlamudi, K., Reddy, P., Adhikari, D., & Liu, K. (2010). Genetically modified mouse models for premature ovarian failure (POF). *Molecular and Cellular Endocrinology*, *315*(1–2), 1–10. https://doi.org/10.1016/j.mce.2009.07.016.

Je, H. D., Gangopadhyay, S. S., Ashworth, T. D., & Morgan, K. G. (2001). Calponin is required for agonist-induced signal transduction—Evidence from an antisense approach in ferret smooth muscle. *Journal of Physiology*, *537*(2), 567–577. https://doi.org/10.1111/j.1469-7793.2001.00567.x.

Jensen, M. H., Morris, E. J., Gallant, C. M., Morgan, K. G., Weitz, D. A., & Moore, J. (2014). Mechanism of calponin stabilization of cross-linked actin networks. *Biophys J*, *106*, 793-800.

Jiang, W. R., Cady, G., Hossain, M. M., Huang, Q. Q., Wang, X., & Jin, J. P. (2014). Mechanoregulation of h2-calponin gene expression and the role of notch signaling. *Journal of Biological Chemistry*, *289*(3), 1617–1628. https://doi.org/10.1074/jbc.M113.498147.

Jin, J. P., Walsh, M. P., Resek, M. E., & McMartin, G. A. (1996). Expression and epitopic conservation of calponin in different smooth muscles and during development. *Biochemistry and Cell Biology = Biochimie Et Biologie Cellulaire*, *74*(2), 187–196. https://doi.org/10.1139/o96-019.

Jin, J. P., Walsh, M. P., Sutherland, C., & Chen, W. (2000). A role for serine-175 in modulating the molecular conformation of calponin. *The Biochemical Journal*, *350 Pt 2*(Pt 2), 579–588.

Jin, J.-P., Zhang, Z., & Bautista, J. A. (2008). Isoform diversity, regulation, and functional adaptation of troponin and calponin. *Critical Reviews in Eukaryotic Gene Expression*, *18*(2), 93–124. https://doi.org/10.1615/critreveukargeneexpr.v18.i2.10.

Jo, S., Kim, H.-R., Mun, Y., & Jun, C.-D. (2018). Transgelin-2 in immunity: Its implication in cell therapy. *Journal of Leukocyte Biology*, *104*(5), 903–910. https://doi.org/10.1002/JLB.MR1117-470R.

Jokubkiene, L., Sladkevicius, P., Rovas, L., & Valentin, L. (2006). Assessment of changes in volume and vascularity of the ovaries during the normal menstrual cycle using three-dimensional power Doppler ultrasound. *Human Reproduction (Oxford, England)*, *21*(10), 2661–2668. https://doi.org/10.1093/humrep/del211.

Jones, G. E. (2000). Cellular signaling in macrophage migration and chemotaxis. *Journal of Leukocyte Biology*, *68*(5), 593–602.

Kageyama, R., Ishibashi, M., Takebayashi, K., & Tomita, K. (1997). bHLH transcription factors and mammalian neuronal differentiation. *Int J Biochem Cell Biol*, *29*, 1389–1399.

Kake, T., Kimura, S., Takahashi, K., & Maruyama, K. (1995). Calponin induces actin polymerization at low ionic strength and inhibits depolymerization of actin filaments. *Biochem J*, *312*(Pt 2), 587–592.

Kamari, Y., Shaish, A., Shemesh, S., Vax, E., Grosskopf, I., Dotan, S., White, M., Voronov, E., Dinarello, C. A., Apte, R. N., & Harats, D. (2011). Reduced atherosclerosis and inflammatory cytokines in apolipoprotein-E-deficient mice lacking bone marrow-derived interleukin-1alpha. *Biochem Biophys Res Commun*, *405*, 197–203.

Kaneko, T., Amano, M., Maeda, A., Goto, H., Takahashi, K., Ito, M., & Kaibuchi, K. (2000). Identification of calponin as a novel substrate of Rho-kinase. *Biochem Biophys Res Commun*, *273*, 110–116.

Kang, J. L., Pack, I. S., Hong, S. M., Lee, H. S., & Castranova, V. (2000). Silica induces nuclear factor-kappa B activation through tyrosine phosphorylation of I kappa B-alpha in RAW264.7 macrophages. *Toxicology and Applied Pharmacology*, *169*(1), 59–65. https://doi.org/10.1006/taap.2000.9039.

Kang, J., Steward, R. L., Kim, Y., Schwartz, R. S., LeDuc, P. R., & Puskar, K. M. (2011). Response of an actin filament network model under cyclic stretching through a coarse grained Monte Carlo approach. *J Theor Biol*, *274*, 109–119.

Kato, R., Hayashi, M., Aiuchi, T., Sawada, N., Obama, T., & Itabe, H. (2019). Temporal and spatial changes of peroxiredoxin 2 levels in aortic media at very early stages of atherosclerotic lesion formation in apoE-knockout mice. *Free Radic Biol Med*, *130*, 348–360.

Kawasaki, H., & Kretsinger, R. H. (2017). Structural and functional diversity of EF-hand proteins: Evolutionary perspectives. *Protein Sci*, *26*(10), 1898–1920.

Kawashima, I., & Kawamura, K. (2018). Regulation of follicle growth through hormonal factors and mechanical cues mediated by Hippo signaling pathway. *Systems Biology in Reproductive Medicine*, *64*(1), 3–11. https://doi.org/10.1080/19396368.2017.1411990.

Khalil, R. A., Menice, C. B., Wang, C. L. A., & Morgan, K. G. (1995). Phosphotyrosine-dependent targeting of mitogen-activated protein kinase in differentiated contractile vascular cells. *Circulation Research*, *76*(6), 1101–1108. https://doi.org/10.1161/01.RES.76.6.1101.

Kim, H. R., Gallant, C., & Morgan, K. G. (2013). Regulation of PKC autophosphorylation by calponin in contractile vascular smooth muscle tissue. *BioMed Research International*. https://doi.org/10.1155/2013/358643.

Kim, M. L., Chae, J. J., Park, Y. H., Nardo, D., Stirzaker, R. A., Ko, H. J., Tye, H., Cengia, L., DiRago, L., Metcalf, D., Roberts, A. W., Kastner, D. L., Lew, A. M., Lyras, D.,

Kile, B. T., Croker, B. A., & Masters, S. L. (2015). Aberrant actin depolymerization triggers the pyrin inflammasome and autoinflammatory disease that is dependent on IL-18, not IL-1beta. *J Exp Med*, *212*(6), 927–938.

Kimura, K., Hashiguchi, T., Deguchi, T., Horinouchi, S., Uto, T., Oku, H., Setoyama, S., Maruyama, I., Osame, M., & Arimura, K. (2007). Serum VEGF–as a prognostic factor of atherosclerosis. *Atherosclerosis*, *194*, 182–188.

Kinne, R. W., Brauer, R., Stuhlmuller, B., Palombo-Kinne, E., & Burmester, G. R. (2000). Macrophages in rheumatoid arthritis. *Arthritis Res*, *2*, 189-202.

Kitching, R., Qi, S., Li, V., Raouf, A., Vary, C. P., & Seth, A. (2002). Coordinate gene expression patterns during osteoblast maturation and retinoic acid treatment of MC3T3-E1 cells. *J Bone Miner Metab*, *20*(5), 269–280.

Kõks, S., Dogan, S., Tuna, B. G., González-Navarro, H., Potter, P., & Vandenbroucke, R. E. (2016). Mouse models of ageing and their relevance to disease. *Mechanisms of Ageing and Development*, *160*, 41–53. https://doi.org/10.1016/j.mad.2016.10.001.

Koninckx, P. R., Gomel, V., Ussia, A., & Adamyan, L. (2016). Role of the peritoneal cavity in the prevention of postoperative adhesions, pain, and fatigue. *Fertil Steril. Oct*, *106*(5), 998–1010. https://doi.org/10.1016/j.fertnstert.2016.08.012.

Kopf, M., Schneider, C., & Nobs, S. P. (2015). The development and function of lung-resident macrophages and dendritic cells. *Nat Immunol*, *16*(1), 36–44.

Kozlowski, J. M., Fidler, I. J., Campbell, D., Xu, Z. L., ME, K., & Hart, I. R. (1984). Metastatic behavior of human tumor cell lines grown in the nude mouse. *Cancer Res*, *44*, 3522–3529.

Kraemer, B., Wallwiener, C., Rajab, T. K., Brochhausen, C., Wallwiener, M., & Rothmund, R. (2014). Standardised models for inducing experimental peritoneal adhesions in female rats. *Biomed Res Int*, *2014*(435056). https://doi.org/10.1155/2014/435056.

Kravić, B., Bionda, T., Siebert, A., Gahlot, P., Levantovsky, S., Behrends, C., & Meyer, H. (2022). Ubiquitin profiling of lysophagy identifies actin stabilizer CNN2 as a target of VCP/p97 and uncovers a link to HSPB1. *Molecular Cell*, *82*(14), 2633-2649 7. https://doi.org/10.1016/j.molcel.2022.06.012.

Kriz, W., & Lemley, K. V. (2015). A potential role for mechanical forces in the detachment of podocytes and the progression of CKD. *J Am Soc Nephrol*, *26*, 258-269.

Kumar, H., Kawai, T., & Akira, S. (2011). Pathogen recognition by the innate immune system. *Int Rev Immunol*, *30*(1), 16–34.

Kumar, T. R., Palapattu, G., Wang, P., Woodruff, T. K., Boime, I., Byrne, M. C., & Matzuk, M. M. (1999). Transgenic models to study gonadotropin function: The role of follicle-stimulating hormone in gonadal growth and tumorigenesis. *Molecular Endocrinology (Baltimore, Md.)*, *13*(6), 851–865. https://doi.org/10.1210/mend.13.6.0297.

Kuo, H. C., Chiu, C. C., & Chang, W. C. (2011). Use of proteomic differential displays to assess functional discrepancies and adjustments of human bone marrow- and Wharton jelly-derived mesenchymal stem cells. *J Proteome Res*, *10*(3), 1305–1315.

Lai, A., Cox, C. D., Chandra Sekar, N., Thurgood, P., Jaworowski, A., Peter, K., & Baratchi, S. (2022). Mechanosensing by Piezo1 and its implications for physiology and various pathologies. *Biol Rev Camb Philos Soc*, *Apr;97(2):604-614*. https://doi.org/10.1111/brv.12814.

Lambrecht, B. N. (2006). Alveolar macrophage in the driver's seat. *Immunity*, *24*(4), 366–368.

Latif, N., Sarathchandra, P., Chester, A. H., & Yacoub, M. H. (2015). Expression of smooth muscle cell markers and co-activators in calcified aortic valves. *Eur Heart J*, *36*(21), 1335–1345.

Lawson, D., Harrison, M., & Shapland, C. (1997). Fibroblast transgelin and smooth muscle SM22alpha are the same protein, the expression of which is down-regulated in many cell lines. *Cell Motil Cytoskeleton*, *38*, 250–257.

Leal, R. F., Milanski, M., Ayrizono Mde, L., Coope, A., Rodrigues, V. S., Portovedo, M., Oliveira, L. M., Fagundes, J. J., Coy, C. S., & Velloso, L. A. (2013). Toll-like receptor 4, F4/80 and pro-inflammatory cytokines in intestinal and mesenteric fat tissue of Crohn's disease. *Int J Clin Exp Med*, *6*(2), 98–104.

Lee, E. H., Han, S. E., Park, M. J., Kim, H. J., Kim, H. G., Kim, C. W., Joo, B. S., & Lee, K. S. (2018). Establishment of Effective Mouse Model of Premature Ovarian Failure Considering Treatment Duration of Anticancer Drugs and Natural Recovery Time. *Journal of Menopausal Medicine*, *24*(3), 196–203. https://doi.org/10.6118/jmm.2018.24.3.196.

Lee, J. H., Perez-Flores, C., Park, S., Kim, H. J., Chen, Y., Kang, M., Kersigo, J., Choi, J., Thai, P., Woltz, R., Perez-Flores, D. C., Perkins, G., Sihn, C. R., Trinh, P., Zhang, X. D., Sirish, P., Dong, Y., Feng, W. W., Pessah, I. N., Dixon, R. E., Sokolowski, B., Fritzsch, B., Chiamvimonvat, N., Yamoah, E. (2024). The Piezo channel is a mechano-sensitive complex component in the mammalian inner ear hair cell. *Nat Commun*. 15(1), 526. doi: 10.1038/s41467-023-44230-x.

Lee, S. H., & Choi, J. H. (2016). Involvement of Immune Cell Network in Aortic Valve Stenosis: Communication between Valvular Interstitial Cells and Immune Cells. *Immune Netw*, *16*(1), 26–32.

Lee, S. Y., Choi, J. Y., Jin, D. C., Kim, J., & Cha, J. H. (2010). Expression of calponin in periglomerular myofibroblasts of rat kidney with experimental chronic injuries. *Anat Cell Biol*, *43*, 132–139.

Lee, Y. G., Lee, J., Byeon, S. E., Yoo, D. S., Kim, M. H., Lee, S. Y., & Cho, J. Y. (2011). Functional role of Akt in macrophage-mediated innate immunity. *Front Biosci*, *16*, 517-530.

Lees-Miller, J. P., Heeley, D. H., & Smillie, L. B. (1987). An abundant and novel protein of 22 kDa (SM22) is widely distributed in smooth muscles. Purification from bovine aorta. *The Biochemical Journal*, *244*(3), 705–709. https://doi.org/10.1042/bj2440705.

Lees-Miller, J. P., Heeley, D. H., Smillie, L. B., & Kay, C. M. (1987). Isolation and characterization of an abundant and novel 22-kDa protein (SM22) from chicken gizzard smooth muscle. *The Journal of Biological Chemistry*, *262*(7), 2988–2993.

Lehoux, S., & Tedgui, A. (2003). Cellular mechanics and gene expression in blood vessels. *Journal of Biomechanics*, *36*(5), 631–643. https://doi.org/10.1016/s0021-9290(02)00441-4.

Leinweber, B. D., Leavis, P. C., Grabarek, Z., Wang, C. L., & Morgan, K. G. (1999). Extracellular regulated kinase (ERK) interaction with actin and the calponin homology (CH) domain of actin-binding proteins. *Biochem J*, *344 Pt 1*, 117–123.

Leinweber, B., Parissenti, A. M., Gallant, C., Gangopadhyay, S. S., Kirwan-Rhude, A., Leavis, P. C., & Morgan, K. G. (2000). Regulation of protein kinase C by the cytoskeletal protein calponin. *Journal of Biological Chemistry*, *275*(51), 40329–40336. https://doi.org/10.1074/jbc.M008257200.

Leinweber, B., Tang, J. X., Stafford, W. F., & Chalovich, J. M. (1999). Calponin interaction with alpha-actinin-actin: Evidence for a structural role for calponin. *Biophys J*, *77*, 3208–3217.

Lerman, D. A., Prasad, S., & Alotti, N. (2015). Calcific Aortic Valve Disease: Molecular Mechanisms and Therapeutic Approaches. *Eur Cardiol*, *10*(2), 108–112.

Li, J., Zhao, Z., Wang, J., Chen, G., Yang, J., & Luo, S. (2008). The role of extracellular matrix, integrins, and cytoskeleton in mechanotransduction of centrifugal loading. *Mol Cell Biochem*, *309*, 41–48.

Li, M., Li, S., Lou, Z., Liao, X., Zhao, X., Meng, Z., Bartlam, M., & Rao, Z. (2008). Crystal structure of human transgelin. *Journal of Structural Biology*, *162*(2), 229–236. https://doi.org/10.1016/j.jsb.2008.01.005.

Li, P., Schwarz, E. M., O'Keefe, R. J., Ma, L., Looney, R. J., Ritchlin, C. T., Boyce, B. F., & Xing, L. (2004). Systemic tumor necrosis factor alpha mediates an increase in peripheral CD11b high osteoclast precursors in tumor necrosis factor alpha-transgenic mice. *Arthritis and Rheumatism*, *50*, 265-276.

Li, X. F., Wang, Y., Zheng, D. D., Xu, H. X., Wang, T., & Pan, M. (2016). M1 macrophages promote aortic valve calcification mediated by microRNA-214/TWIST1 pathway in valvular interstitial cells. *Am J Transl Res*, *8*(12), 5773–5783.

Li, Y., Lee, P. Y., & Reeves, W. H. (2010). Monocyte and macrophage abnormalities in systemic lupus erythematosus. *Arch Immunol Ther Exp (Warsz*, *58*, 355-364.

Liakakos, T., Thomakos, N., Fine, P. M., Dervenis, C., & Young, R. L. (2001). Peritoneal adhesions: Etiology, pathophysiology, and clinical significance. Recent advances in prevention and management. *Dig Surg*, *18*(4), 260–273. https://doi.org/10.1159/000050149.

Libby, P. (2012). Inflammation in atherosclerosis. *Arterioscler Thromb Vasc Biol*, *32* (9), 2045-51.

Lim, C. G., Jang, J., & Kim, C. (2018). Cellular machinery for sensing mechanical force. *BMB Reports*, *51*(12), 623–629. https://doi.org/10.5483/BMBRep.2018.51.12.237.

Lim, S. M., Trzeciakowski, J. P., Sreenivasappa, H., Dangott, L. J., & Trache, A. (2012). RhoA-induced cytoskeletal tension controls adaptive cellular remodeling to mechanical signaling. *Integrative Biology (United Kingdom*, *4*(6), 615–627. https://doi.org/10.1039/c2ib20008b

Lin, J. J., Li, Y., Eppinga, R. D., Wang, Q., & Jin, J.-P. (2009). Chapter 1: Roles of caldesmon in cell motility and actin cytoskeleton remodeling. *Int. Rev. Cell Mol. Biol*, *274*, 1–68.

Liu, A. C., Joag, V. R., & Gotlieb, A. I. (2007). The emerging role of valve interstitial cell phenotypes in regulating heart valve pathobiology. *Am J Pathol*, *171*(5), 1407–1418.

Liu, G., Wu, C., Wu, Y., & Zhao, Y. (2006). Phagocytosis of apoptotic cells and immune regulation. *Scandinavian journal of immunology*, *64*, 1-9.

Liu, H., Shi, B., Huang, Q. Q., Eksarko, P., & Pope, R. M. (2008). Transcriptional diversity during monocyte to macrophage differentiation. *Immunol Lett*, *117*, 70-80.

Liu, J., Zhang, Y., Li, Q., & Wang, Y. (2020). Transgelins: Cytoskeletal Associated Proteins Implicated in the Metastasis of Colorectal Cancer. *Front Cell Dev Biol, 8*, 573859. doi: 10.3389/fcell.2020.573859.

Liu, L., Meng, T., Zheng, X., Liu, Y., Hao, R., Yan, Y., Chen, S., You, H., Xing, J., & Dong, Y. (2019). Transgelin 2 Promotes Paclitaxel Resistance, Migration, and Invasion of Breast Cancer by Directly Interacting with PTEN and Activating PI3K/Akt/GSK-3β Pathway. *Molecular Cancer Therapeutics, 18*(12), 2457–2468. https://doi.org/10.1158/1535-7163.MCT-19-0261.

Liu, R., Hossain, M. M., Chen, X., & Jin, J. P. (2017). Mechanoregulation of SM22α/Transgelin. *Biochemistry, 56*(41), 5526–5538. https://doi.org/10.1021/acs.biochem.7b00794.

Liu, R., & Jin, J. P. (2016a). Calponin isoforms CNN1, CNN2 and CNN3: Regulators for actin cytoskeleton functions in smooth muscle and non-muscle cells. *Gene, 585*(1), 143–153. https://doi.org/10.1016/j.gene.2016.02.040.

Liu, R., & Jin, J. P. (2016b). Deletion of calponin 2 in macrophages alters cytoskeleton-based functions and attenuates the development of atherosclerosis. *Journal of Molecular and Cellular Cardiology, 99*, 87–99. https://doi.org/10.1016/j.yjmcc.2016.08.019.

Liu, Y. C., Zou, X. B., Chai, Y. F., & Yao, Y. M. (2014). Macrophage polarization in inflammatory diseases. *Int J Biol Sci, 10*(5), 520–529.

Long, X., Slivano, O. J., Cowan, S. L., Georger, M. A., Lee, T. H., & Miano, J. M. (2011). Smooth muscle calponin: An unconventional CArG-dependent gene that antagonizes neointimal formation. *Arterioscler Thromb Vasc Biol, 31*, 2172–2180.

Ludewig, B., & Laman, J. D. (2004). The in and out of monocytes in atherosclerotic plaques: Balancing inflammation through migration. *Proc Natl Acad Sci USA, 101*, 11529–11530.

Lundahl, J., Hallden, G., & Skold, C. M. (1996). Human blood monocytes, but not alveolar macrophages, reveal increased CD11b/CD18 expression and adhesion properties upon receptor-dependent activation. *Eur Respir J, 9*(6), 1188–1194.

Lusis, A. J. (2000). Atherosclerosis. *Nature, 407*, 233–241.

Ma, H., Killaars, A. R., DelRio, F. W., Yang, C., & Anseth, K. S. (2017). Myofibroblastic activation of valvular interstitial cells is modulated by spatial variations in matrix elasticity and its organization. *Biomaterials, 131*, 131–144.

Ma, L. J., & Fogo, A. B. (2003). Model of robust induction of glomerulosclerosis in mice: Importance of genetic background. *Kidney Int, 64*, 350-355.

Ma, Y., Bogatcheva, N. V., & Gusev, N. B. (2000). Heat shock protein (hsp90) interacts with smooth muscle calponin and affects calponin-binding to actin. *Biochim Biophys Acta, 1476*(2), 300–310.

Mabuchi, K., Li, B., Ip, W., & Tao, T. (1997). Association of calponin with desmin intermediate filaments. *The Journal of Biological Chemistry, 272*, 22662-22666.

Maccioni, M., Zeder-Lutz, G., Huang, H., Ebel, C., Gerber, P., Hergueux, J., Marchal, P., Duchatelle, V., Degott, C., Regenmortel, M., Benoist, C., & Mathis, D. (2002). Arthritogenic monoclonal antibodies from K/BxN mice. *The Journal of Experimental Medicine, 195*, 1071-1077.

Maguchi, M., Nishida, W., Kohara, K., Kuwano, A., Kondo, I., & Hiwada, K. (1995). Molecular cloning and gene mapping of human basic and acidic calponins. *Biochem Biophys Res Commun*, *217*, 238–244.

Makuch, R., Birukov, K., Shirinsky, V., & Dabrowska, R. (1991). Functional interrelationship between calponin and caldesmon. *Biochem J*, *280*(Pt 1), 33–38. https://doi.org/10.1042/bj2800033.

Man, S. M., Ekpenyong, A., Tourlomousis, P., Achouri, S., Cammarota, E., Hughes, K., Rizzo, A., Ng, G., Wright, J. A., Cicuta, P., Guck, B., JR, & C.E. (2014). Actin polymerization as a key innate immune effector mechanism to control Salmonella infection. *Proceedings of the National Academy of Sciences*, *111*(49), 17588–17593.

Martin, D. K., Bootcov, M. R., Campbell, T. J., French, P. W., & Breit, S. N. (1995). Human macrophages contain a stretch-sensitive potassium channel that is activated by adherence and cytokines. *J Membr Biol*, *147*(3), 305–315.

Martín de las Mulas, J., Reymundo, C., Espinosa de los Monteros, A., Millán, Y., & Ordás, J. (2004). Calponin expression and myoepithelial cell differentiation in canine, feline and human mammary simple carcinomas. *Veterinary and Comparative Oncology*, *2*(1), 24–35. https://doi.org/10.1111/j.1476-5810.2004.00036.x.

Maselli, V., Galdiero, E., Salzano, A. M., Scaloni, A., Maione, A., Falanga, A., Naviglio, D., Guida, M., Di Cosmo, A., & Galdiero, S. (2020). OctoPartenopin: Identification and Preliminary Characterization of a Novel Antimicrobial Peptide from the Suckers of Octopus vulgaris. *Mar Drugs*, *18*(8), 380. doi: 10.3390/md18080380.

Masuki, S., Takeoka, M., Taniguchi, S., & Nose, H. (2003). Enhanced baroreflex sensitivity in free-moving calponin knockout mice. *American Journal of Physiology. Heart and Circulatory Physiology*, *284*(3), H939-946. https://doi.org/10.1152/ajpheart.00610.2002.

Masuki, S., Takeoka, M., Taniguchi, S., Yokoyama, M., & Nose, H. (2003). Impaired arterial pressure regulation during exercise due to enhanced muscular vasodilatation in calponin knockout mice. *The Journal of Physiology*, *553*(Pt 1), 203–212. https://doi.org/10.1113/jphysiol.2003.047803.

Masur, S. K., Dewal, H. S., Dinh, T. T., Erenburg, I., & Petridou, S. (1996). Myofibroblasts differentiate from fibroblasts when plated at low density. *Proc Natl Acad Sci U S A*, *30;93(9):4219-23.*

Matthew, J. D., Khromov, A. S., McDuffie, M. J., Somlyo, A. V., Somlyo, A. P., Taniguchi, S., & Takahashi, K. (2000). Contractile properties and proteins of smooth muscles of a calponin knockout mouse. *Journal of Physiology*, *529*(3), 811–824. https://doi.org/10.1111/j.1469-7793.2000.00811.x

Matthews, B. D., Overby, D. R., Mannix, R., & Ingber, D. E. (2006). Cellular adaptation to mechanical stress: Role of integrins, Rho, cytoskeletal tension and mechanosensitive ion channels. *Journal of Cell Science*, *119*(3), 508–518. https://doi.org/10.1242/jcs.02760.

May, R. C., & Machesky, L. M. (2001). Phagocytosis and the actin cytoskeleton. *J Cell Sci*, *114*(Pt 6), 1061–1077.

Mazaud Guittot, S., Guigon, C. J., Coudouel, N., & Magre, S. (2006). Consequences of fetal irradiation on follicle histogenesis and early follicle development in rat ovaries.

Biology of Reproduction, *75*(5), 749–759. https://doi.org/10.1095/biolreprod.105.050633.

McWhorter, F. Y., Davis, C. T., & Liu, W. F. (2015). Physical and mechanical regulation of macrophage phenotype and function. *Cell Mol Life Sci*, *72*(7), 1303–1316.

McWhorter, F. Y., Wang, T., Nguyen, P., Chung, T., & Liu, W. F. (2013). Modulation of macrophage phenotype by cell shape. *Proc Natl Acad Sci U S A*, *110*(43), 17253–17258.

Meehan, K. L., Holland, J. W., & Dawkins, H. J. (2002). Proteomic analysis of normal and malignant prostate tissue to identify novel proteins lost in cancer. *Prostate*, *50*, 54–63.

Meng, T., Liu, L., Hao, R., Chen, S., & Dong, Y. (2017). Transgelin-2: A potential oncogenic factor. *Tumour Biol*, *39*(6), 1010428317702650. doi: 10.1177/1010428317702650.

Menice, C. B., Hulvershorn, J., Adam, L. P., Wang, C. A., & Morgan, K. G. (1997). Calponin and mitogen-activated protein kinase signaling in differentiated vascular smooth muscle. *J Biol Chem*, *272*(40), 25157–25161.

Merryman, W. D., Bieniek, P. D., Guilak, F., & Sacks, M. S. (2009). Viscoelastic properties of the aortic valve interstitial cell. *J Biomech Eng*, *131*(4), 041005. doi: 10.1115/1.3049821.

Merryman, W. D., Youn, I., Lukoff, H. D., Krueger, P. M., Guilak, F., Hopkins, R. A., & Sacks, M.S. (2006). Correlation between heart valve interstitial cell stiffness and transvalvular pressure: Implications for collagen biosynthesis. *Am J Physiol Heart Circ Physiol*, *290*(1), H224-31.

Meyer, T., Unterberg, C., Kreuzer, H., & Buchwald, A. B. (1996). Accumulation of unphosphorylated calponin in the submembranous cytoskeletons of arachidonic acid-stimulated human platelets. *Thromb Haemost*, *75*, 617-622.

Meyer-Rochow, V. B., Fraile, B., Paniagua, R., & Royuela, M. (2003). First immunocytochemical study of echinoderm smooth muscle: The Antarctic cushionstar Odontaster validus Koehler (Echinodermata, Asteroidea. *Protoplasma*, *220*(3–4), 227–232.

Meyrelles, S. S., Peotta, V. A., Pereira, T. M., & Vasquez, E. C. (2011). Endothelial dysfunction in the apolipoprotein E-deficient mouse: Insights into the influence of diet, gender and aging. *Lipids Health Dis*, *10*, 211. doi: 10.1186/1476-511X-10-211.

Mezgueldi, M., Fattoum, A., Derancourt, J., & Kassab, R. (1992). Mapping of the functional domains in the amino-terminal region of calponin. *J Biol Chem*, *267*(22), 15943–15951.

Mezgueldi, M., Mendre, C., Calas, B., Kassab, R., & Fattoum, A. (1995). Characterization of the regulatory domain of gizzard calponin. Interactions of the 145-163 region with F-actin, calcium-binding proteins, and tropomyosin. *J Biol Chem*, *270*, 8867–8876.

Miano, J. M., Krahe, R., Garcia, E., Elliott, J. M., & Olson, E. N. (1997). Expression, genomic structure and high resolution mapping to 19p13.2 of the human smooth muscle cell calponin gene. *Gene*, *15;197(1-2):215-24*. https://doi.org/10.1016/s0378-1119(97)00265-5.

Miller, J. D., Weiss, R. M., & Heistad, D. D. (2011). Calcific aortic valve stenosis: Methods, models, and mechanisms. *Circ Res*, *108*(11), 1392–1412.

Mino, T., Yuasa, U., Nakamura, F., Naka, M., & Tanaka, T. (1998). Two distinct actin-binding sites of smooth muscle calponin. *Eur J Biochem*, *251*, 262–268.

Minton, K. (2017). Immunometabolism: Stress-induced macrophage polarization. *Nat Rev Immunol*, *17*(5), 277. doi: 10.1038/nri.2017.41.

Mitsios, J. V., Prevost, N., Kasirer-Friede, A., Gutierrez, E., Groisman, A., Abrams, C. S., Wang, Y., Litvinov, R. I., Zemljic-Harpf, A., Ross, R. S., & Shattil, S. J. (2010). What is vinculin needed for in platelets? *J Thromb Haemost*, *8*, 2294-2304.

Moazzem Hossain, M., Wang, X., Bergan, R. C., & Jin, J.-P. (2014). Diminished expression of h2-calponin in prostate cancer cells promotes cell proliferation, migration and the dependence of cell adhesion on substrate stiffness. *FEBS Open Bio*, *4*(1), 627–636. https://doi.org/10.1016/j.fob.2014.06.003.

Mohler, E. R., Gannon, F., Reynolds, C., Zimmerman, R., Keane, M. G., & Kaplan, F. S. (2001). Bone formation and inflammation in cardiac valves. *Circulation*, *103*(11), 1522–1528. https://doi.org/10.1161/01.cir.103.11.1522.

Monach, P. A., Mathis, D., & Benoist, C. (2008). The K/BxN arthritis model. *Curr Protoc Immunol Chapter*, *15: Unit 15*, 22.

Monach, P. A., Nigrovic, P. A., Chen, M., Hock, H., Lee, D. M., Benoist, C., & Mathis, D. (2010). Neutrophils in a mouse model of autoantibody-mediated arthritis: Critical producers of Fc receptor gamma, the receptor for C5a, and lymphocyte function-associated antigen 1. *Arthritis and Rheumatism*, *62*, 753-764,.

Moore, K. J., Sheedy, F. J., & Fisher, E. A. (2013). Macrophages in atherosclerosis: A dynamic balance. *Nat Rev Immunol*, *13*, 709–721.

Morachevskaya, E. A. & Sudarikova, A. V. (2021) Actin dynamics as critical ion channel regulator: ENaC and Piezo in focus. *Am J Physiol Cell Physiol*. 320(5), C696-C702.

Morgan, K. G., & Gangopadhyay, S. S. (2001). Invited review: Cross-bridge regulation by thin filament-associated proteins. *Journal of Applied Physiology (Bethesda, Md.: 1985)*, *91*(2), 953–962. https://doi.org/10.1152/jappl.2001.91.2.953.

Morgan, R., Hooiveld, M. H., Pannese, M., Dati, G., Broders, F., Delarue, M., Thiery, J. P., Boncinelli, E., & Durston, A. J. (1999). Calponin modulates the exclusion of Otx-expressing cells from convergence extension movements. *Nat Cell Biol*, *1*, 404–408.

Morioka, T., Koyama, H., Yamamura, H., Tanaka, S., Fukumoto, S., Emoto, M., Mizuguchi, H., Hayakawa, T., Kojima, I., Takahashi, K., & Nishizawa, Y. (2003). Role of H1-calponin in pancreatic AR42J cell differentiation into insulin-producing cells. *Diabetes*, *52*, 760–766.

Morrow, D., Sweeney, C., Birney, Y. A., Cummins, P. M., Walls, D., Redmond, E. M., & Cahill, P. A. (2005). Cyclic strain inhibits Notch receptor signaling in vascular smooth muscle cells in vitro. *Circ Res*, *96*, 567–575.

Morrow, D., Sweeney, C., Birney, Y. A., Guha, S., Collins, N., Cummins, P. M., Murphy, R., Walls, D., Redmond, E. M., & Cahill, P. A. (2007). Biomechanical regulation of hedgehog signaling in vascular smooth muscle cells in vitro and in vivo. *Am J Physiol Cell Physiol*, *292*, 488–496.

Moscat, J., & Diaz-Meco, M. T. (2009). To aggregate or not to aggregate? A new role for p62. *EMBO Rep*, *10*(8), 804. doi: 10.1038/embor.2009.172.

Mostowy, S., & Shenoy, A. R. (2015). The cytoskeleton in cell-autonomous immunity: Structural determinants of host defence. *Nat Rev Immunol*, *15*(9), 559–573.

Mosunjac, M. B., Lewis, M. M., Lawson, D., & Cohen, C. (2000). Use of a novel marker, calponin, for myoepithelial cells in fine-needle aspirates of papillary breast lesions. *Diagn Cytopathol*, *23*, 151–155.

Mozaffarian, D., Benjamin, E. J., Go, A. S., Arnett, D. K., & Blaha, M. J. (2016). Executive Summary: Heart Disease and Stroke Statistics–2016 Update: A Report From the American Heart Association. *Circulation*, *133*(4), 447–454.

Mukherjee, P., Rahaman, S. G., Goswami, R., Dutta, B., Mahanty, M., & Rahaman, S. O. (2022). Role of mechanosensitive channels/receptors in atherosclerosis. *Am J Physiol Cell Physiol*, *322*(5), 927–938.

Murashita, T., & Badhwar, V. (2017). In search of the smoking gun in calcific aortic valve disease. *J Thorac Cardiovasc Surg*, *153*(6), 1328–1329.

Mutsaers, S. E., Prele, C. M., Pengelly, S., & Herrick, S. E. (2016). Mesothelial cells and peritoneal homeostasis. *Fertil Steril. Oct*, *106*(5), 1018–1024. https://doi.org/10.1016/j.fertnstert.2016.09.005.

Na, B. R., Kim, H. R., & Piragyte, I. (2015). TAGLN2 regulates T cell activation by stabilizing the actin cytoskeleton at the immunological synapse. *J Cell Biol*, *209*(1), 143–162.

Nagamatsu, G., Shimamoto, S., Hamazaki, N., Nishimura, Y., & Hayashi, K. (2019). Mechanical stress accompanied with nuclear rotation is involved in the dormant state of mouse oocytes. *Sci Adv*, *5*(6). https://www.science.org/doi/10.1126/sciadv.aav9960

Nahrendorf, M., & Swirski, F. K. (2015). Immunology. Neutrophil-macrophage communication in inflammation and atherosclerosis. *Science*, *349*, 237–238.

Naka, M., Kureishi, Y., Muroga, Y., Takahashi, K., Ito, M., & Tanaka, T. (1990). Modulation of smooth muscle calponin by protein kinase C and calmodulin. *Biochemical and Biophysical Research Communications*, *171*(3), 933–937. doi.org/10.1016/0006-291x(90)90773-g.

Nakano, K., Bunai, F., & Numata, O. (2005). Stg 1 is a novel SM22/transgelin-like actin-modulating protein in fission yeast. *FEBS Lett*, *579*(28), 6311–6316.

Namvar, S., Woolf, A. S., Zeef, L. A., Wilm, T., Wilm, B., & Herrick, S. E. (2018). Functional molecules in mesothelial-to-mesenchymal transition revealed by transcriptome analyses. *J Pathol. Aug*, *245*(4), 491–501. https://doi.org/10.1002/path.5101.

Nanavati, B. N., Noordstra, I., Verma, S., Duszyc, K., Green, K. J., & Yap, A. S. (2023). *Desmosome-anchored intermediate filaments facilitate tension-sensitive RhoA signaling for epithelial homeostasis*. The Preprint Server for Biology. https://doi.org/10.1101/2023.02.23.529786.

Nasir, K., Katz, R., Takasu, J., Shavelle, D. M., Detrano, R., & Lima, J. A. (2008). Ethnic differences between extra-coronary measures on cardiac computed tomography: Multi-ethnic study of atherosclerosis (MESA. *Atherosclerosis*, *198*(1), 104–114.

Neeves, K. B., Maloney, S. F., Fong, K. P., Schmaier, A. A., Kahn, M. L., Brass, L. F., & Diamond, S. L. (2008). Microfluidic focal thrombosis model for measuring murine platelet deposition and stability: PAR4 signaling enhances shear-resistance of platelet aggregates. *J Thromb Haemost*, *6*, 2193-2201.

Nesbitt, W. S., Westein, E., Tovar-Lopez, F. J., Tolouei, E., Mitchell, A., Fu, J., Carberry, J., Fouras, A., & Jackson, S. P. (2009). A shear gradient-dependent platelet aggregation mechanism drives thrombus formation. *Nat Med*, *15*, 665-673,.

Nigam, R., Triggle, C. R., & Jin, J. P. (1998). And h2-calponins are not essential for norepinephrine- or sodium fluoride-induced contraction of rat aortic smooth muscle. *Journal of Muscle Research and Cell Motility*, *19*(6), 1-. https://doi.org/10.1023/a:1005389300151.

Nishimura, R. A., Otto, C. M., Bonow, R. O., Carabello, B. A., Erwin, J. P., Fleisher, L. A., Jneid, H., Mack, M. J., McLeod, C. J., O'Gara, P. T., Rigolin, V. H., Sundt, T. M., & Thompson, A. (2017). 2017 AHA/ACC Focused Update of the 2014 AHA/ACC Guideline for the Management of Patients with Valvular Heart Disease: A Report of the American College of Cardiology/American Heart Association Task Force on Clinical Practice Guidelines. *Journal of the American College of Cardiology*, *70*(2), 252–289. https://doi.org/10.1016/j.jacc.2017.03.011.

North, A. J., Gimona, M., Cross, R. A., & Small, J. V. (1994). Calponin is localised in both the contractile apparatus and the cytoskeleton of smooth muscle cells. *Journal of Cell Science*, *107*(3), 437–444. https://doi.org/10.1242/jcs.107.3.437.

O'Brien, K. D. (2006). Pathogenesis of calcific aortic valve disease: A disease process comes of age (and a good deal more. *Arterioscler Thromb Vasc Biol*, *26*(8), 1721–1728.

Ohashi, R., Mu, H., Wang, X., Yao, Q., & Chen, C. (2005). Reverse cholesterol transport and cholesterol efflux in atherosclerosis. *QJM*, *98*, 845–856.

Okamoto-Inoue, M., Kamada, S., G, K., & Taniguchi, S. (1999). The induction of smooth muscle alpha actin in a transformed rat cell line suppresses malignant properties in vitro and in vivo. *Cancer Lett*, *142*, 173–178.

Ono, K., Obinata, T., Yamashiro, S., Liu, Z., & Ono, S. (2015). UNC-87 isoforms, Caenorhabditis elegans calponin-related proteins, interact with both actin and myosin and regulate actomyosin contractility. *Mol Biol Cell*, *26*(9), 1687–1698.

Ono, S. (2021). Diversification of the calponin family proteins by gene amplification and repeat expansion of calponin-like motifs. *Cytoskeleton (Hoboken*, *78*(5), 199–205.

Oonuma, K., Yamamoto, M., & Moritsugu, N. (2021). Evolution of Developmental Programs for the Midline Structures in Chordates: Insights From Gene Regulation in the Floor Plate and Hypochord Homologues of Ciona Embryos. *Front Cell Dev Biol*, *9*(704367).

Otto, C. M. (2008). Calcific aortic stenosis–time to look more closely at the valve. *N Engl J Med*, *359*(13), 1395–1398.

Otto, C. M., Kuusisto, J., Reichenbach, D. D., Gown, A. M., & O'Brien, K. D. (1994). Characterization of the early lesion of "degenerative" valvular aortic stenosis. *Histological and Immunohistochemical Studies. Circulation*, *90*(2), 844–853.

Owens, G. K. (1998). Molecular control of vascular smooth muscle cell differentiation. *Acta Physiol Scand*, *164*, 623–635.

Paino, I. M., Miranda, J. C., Marzocchi-Machado, C. M., Cesarino, E. J., Castro, F. A., & Souza, A. M. (2011). Phagocytosis and nitric oxide levels in rheumatic inflammatory states in elderly women. *J Clin Lab Anal*, *25*, 47-51.

Panasenko, O. O., & Gusev, N. B. (2001). Mutual effects of alpha-actinin, calponin and filamin on actin binding. *Biochim Biophys Acta*, *1544*, 393-405.

Parker, C. A., Takahashi, K., Tang, J. X., Tao, T., & Morgan, K. G. (1998). Cytoskeletal targeting of calponin in differentiated, contractile smooth muscle cells of the ferret. *J Physiol*, *508*(Pt 1), 187–198.

Parker, C. A., Takahashi, K., Tao, T., & Morgan, K. G. (1994). Agonist-induced redistribution of calponin in contractile vascular smooth muscle cells. *Am J Physiol*, *267*, 1262–1270. https://doi.org/10.1152/ajpcell.1994.267.5.c1262.

Pataki, M., Lusztig, G., & Robenek, H. (1992). Endocytosis of oxidized LDL and reversibility of migration inhibition in macrophage-derived foam cells in vitro. A mechanism for atherosclerosis regression? *Arterioscler Thromb*, *12*, 936–944.

Patel, N. R., Bole, M., Chen, C., Hardin, C. C., Kho, A. T., Mih, J., Deng, L., Butler, J., Tschumperlin, D., Fredberg, J. J., Krishnan, R., & Koziel, H. (2012). Cell elasticity determines macrophage function. *PLoS One*, *7*, 41024.

Pearlstone, J. R., Weber, M., & Lees-Miller, J. P. (1987). Amino acid sequence of chicken gizzard smooth muscle SM22 alpha. *J Biol Chem*, *262*, 5985–5991.

Pennacchio, F. A., Nastały, P., Poli, A., & Maiuri, P. (2021). Tailoring Cellular Function: The Contribution of the Nucleus in Mechanotransduction. *Frontiers in Bioengineering and Biotechnology*, *8*(January), 1–15. https://doi.org/10.3389/fbioe.2020.596746.

Pergola, C., Schubert, K., Pace, S., Ziereisen, J., Nikels, F., Scherer, O., Hüttel, S., Zahler, S., Vollmar, A. M., Weinigel, C., Rummler, S., Müller, R., Raasch, M., Mosig, A., Koeberle, A., & Werz, O. (2017). Modulation of actin dynamics as potential macrophage subtype-targeting anti-tumour strategy. *Scientific Reports*, *7*(41434).

Pertuy, F., Eckly, A., Weber, J., Proamer, F., Rinckel, J. Y., Lanza, F., Gachet, C., & Leon, C. (2014). Myosin IIA is critical for organelle distribution and F-actin organization in megakaryocytes and platelets. *Blood,* 123(8), 1261-9.

Pfuhl, M., Al-Sarayreh, S., & El-Mezgueldi, M. (2011). The calponin regulatory region is intrinsically unstructured: Novel insight into actin-calponin and calmodulin-calponin interfaces using NMR spectroscopy. *Biophysical Journal*, *100*(7), 1718–1728. https://doi.org/10.1016/j.bpj.2011.01.040.

Plantier, M., Fattoum, A., Menn, B., Ben-Ari, Y., Terrossian, E., & Represa, A. (1999). Acidic calponin immunoreactivity in postnatal rat brain and cultures: Subcellular localization in growth cones, under the plasma membrane and along actin and glial filaments. *European Journal of Neuroscience*, *11*(8), 2801–2812. https://doi.org/10.1046/j.1460-9568.1999.00702.x.

Plazyo, O., Liu, R., Moazzem Hossain, M., & Jin, J. P. (2018). Deletion of calponin 2 attenuates the development of calcific aortic valve disease in ApoE−/− mice. *Journal of Molecular and Cellular Cardiology*, *121*(May), 233–241. https://doi.org/10.1016/j.yjmcc.2018.07.249.

Plazyo, O., Sheng, J. J., & Jin, J. P. (2019). Downregulation of calponin 2 contributes to the quiescence of lung macrophages. *American Journal of Physiology - Cell Physiology*, *317*(4), 749–761. https://doi.org/10.1152/ajpcell.00036.2019.

Plesch, B. (1977). An ultrastructural study of the musculature of the pond snail Lymnaea stagnalis (l. *Cell Tissue Res*, *180*(3), 317–340.

Polte, T. R., Eichler, G. S., N, W., & Ingber, D. E. (2004). Extracellular matrix controls myosin light chain phosphorylation and cell contractility through modulation of cell shape and cytoskeletal prestress. *Am J Physiol Cell Physiol*, *286*, 518–528.

Ponti, A., Machacek, M., Gupton, S. L., CM, W.-S., & Danuser, G. (2004). Two distinct actin networks drive the protrusion of migrating cells. *Science*, *305*, 1782–1786.

Popat, A., Patel, A. A., & Warnes, G. (2018). A Flow Cytometric Study of ER Stress and Autophagy. *Cytometry A*, 95(6), 672-682.

Porras, A. M., Engeland, N. C., Marchbanks, E., McCormack, A., Bouten, C. V., & Yacoub, M. H. (2017). Robust Generation of Quiescent Porcine Valvular Interstitial Cell Cultures. *J Am Heart Assoc*, *6*(3), e005041. doi: 10.1161/JAHA.116.005041.

Prasad, P. D., Stanton, J. A., & Assinder, S. J. (2010). Expression of the actin-associated protein transgelin (SM22) is decreased in prostate cancer. *Cell Tissue Res*, *339*, 337–347.

Previtera, M. L., & Sengupta, A. (2015). Substrate Stiffness Regulates Proinflammatory Mediator Production through TLR4 Activity in Macrophages. *PLoS One*, *10*, 0145813.

Prinjha, R. K., Shapland, C. E., & Hsuan, J. J. (1994). Cloning and sequencing of cDNAs encoding the actin cross-linking protein transgelin defines a new family of actin-associated proteins. *Cell Motil Cytoskeleton*, *28*, 243–255.

Prudovsky, I., Popov, K., Akimov, S., Serov, S., Zelenin, A., Meinhardt, G., Baier, P., Sohn, C., & Hass, R. (2002). Antisense CD11b integrin inhibits the development of a differentiated monocyte/macrophage phenotype in human leukemia cells. *European Journal of Cell Biology*, *81*, 36-42.

Qian, A., Hsieh, T. B., Hossain, M. M., Lin, J. J. C., & Jin, J. P. (2021). A rapid degradation of calponin 2 is required for cytokinesis. *American Journal of Physiology - Cell Physiology*, *321*(2), 355–368. https://doi.org/10.1152/ajpcell.00569.2020.

Qiao, Y. N., He, W. Q., Chen, C. P., Zhang, C. H., Zhao, W., Wang, P., Zhang, L., Wu, Y. Z., Yang, X., Peng, Y. J., Gao, J. M., Kamm, K. E., Stull, J. T., & Zhu, M. S. (2014). Myosin phosphatase target subunit 1 (MYPT1) regulates the contraction and relaxation of vascular smooth muscle and maintains blood pressure. *J Biol Chem*, *289*(32), 22512–22523.

Qin, C., Nagao, T., Grosheva, I., Maxfield, F. R., & Pierini, L. M. (2006). Elevated plasma membrane cholesterol content alters macrophage signaling and function. *Arterioscler Thromb Vasc Biol*, *26*, 372–378.

Qiu, Z., Chu, Y., Xu, B., Wang, Q., Jiang, M., Li, X., Wang, G., Yu, P., Liu, G., Wang, H., Kang, H., Liu, J., Zhang, Y., Jin, J. P., Wu, K., & Liang, J. (2017). Increased expression of calponin 2 is a positive prognostic factor in pancreatic ductal adenocarcinoma. *Oncotarget*, *8*(34), 56428–56442. https://doi.org/10.18632/oncotarget.17701.

Rabkin-Aikawa, E., Aikawa, M., Farber, M., Kratz, J. R., Garcia-Cardena, G., Kouchoukos, N. T., Mitchell, M. B., Jonas, R. A., & Schoen, F. J. (2004). Clinical pulmonary autograft valves: Pathologic evidence of adaptive remodeling in the aortic site. *The Journal of Thoracic and Cardiovascular Surgery*, *128*(4), 552–561. https://doi.org/10.1016/j.jtcvs.2004.04.016

Rabkin-Aikawa, E., Farber, M., Aikawa, M., & Schoen, F. J. (2004). Dynamic and reversible changes of interstitial cell phenotype during remodeling of cardiac valves. *The Journal of Heart Valve Disease*, *13*(5), 841–847.

Rajamannan, N. M., Evans, F. J., Aikawa, E., Grande-Allen, K. J., Demer, L. L., & Heistad, D. D. (2011). Calcific aortic valve disease: Not simply a degenerative process: A review and agenda for research from the National Heart and Lung and Blood Institute Aortic Stenosis Working Group. Executive summary: Calcific aortic valve disease-2011 update. *Circulation*, *124*(16), 1783–1791.

Rajamannan, N. M., Subramaniam, M., Rickard, D., Stock, D., SR, J, S., & M. (2003). Human aortic valve calcification is associated with an osteoblast phenotype. *Circulation*, *107*(17), 2181–2184.

Rami, G., Caillard, O., Medina, I., Pellegrino, C., Fattoum, A., Ben-Ari, Y., & Ferhat, L. (2006). Change in the shape and density of dendritic spines caused by overexpression of acidic calponin in cultured hippocampal neurons. *Hippocampus*, *16*, 183–197.

Ramji, D. P., & Davies, T. S. (2015). Cytokines in atherosclerosis: Key players in all stages of disease and promising therapeutic targets. *Cytokine Growth Factor Rev*, *26*, 673–685.

Rana, N., Braun, D. P., House, R., Gebel, H., Rotman, C., & Dmowski, W. P. (1996). Basal and stimulated secretion of cytokines by peritoneal macrophages in women with endometriosis. *Fertil Steril*, *65*(5), 925–930.

Rasmussen, M., & Jin, J.-P. (2024). Mechanoregulation and Function of Calponin and Transgelin. *Biophysics Reviews*, *5*(1). https://doi.org/10.1063/5.0176784.

Rath, M., Muller, I., Kropf, P., Closs, E. I., & Munder, M. (2014). Metabolism via Arginase or Nitric Oxide Synthase: Two Competing Arginine Pathways in Macrophages. *Front Immunol*, *5*, 532. https://doi.org/10.3389/fimmu.2014.00532

Rattazzi, M., & Pauletto, P. (2015). Valvular endothelial cells: Guardians or destroyers of aortic valve integrity? *Atherosclerosis*, *242*(2), 396–398.

Reichl, E. M., Effler, J. C., & Robinson, D. N. (2005). The stress and strain of cytokinesis. *Trends Cell Biol*, *15*(4), 200–206. https://doi.org/10.1016/j.tcb.2005.02.004.

Ren, W. Z., Ng, G. Y., & Wang, R. X. (1994). The identification of NP25: A novel protein that is differentially expressed by neuronal subpopulations. *Brain Res Mol Brain Res*, *22*(1–4), 173–185.

Represa, A., Trabelsi-Terzidis, H., Plantier, M., Fattoum, A., Jorquera, I., Agassandian, C., Ben-Ari, Y., & Terrossian, E. (1995). Distribution of caldesmon and of the acidic isoform of calponin in cultured cerebellar neurons and in different regions of the rat brain: An immunofluorescence and confocal microscopy study. *Experimental Cell Research*, *221*(2), 333–343. https://doi.org/10.1006/excr.1995.1383.

Reverendo, M., Mendes, A., Arguello, R. J., Gatti, E., & Pierre, P. (2019). At the crossway of ER-stress and proinflammatory responses. *FEBS J*, *286*(2), 297–310.

Rinaldi, G. J. (2005). Blood pressure fall and increased relaxation of aortic smooth muscle in diabetic rats. *Diabetes Metab*, *31*(5), 487–495.

Ringvold, H. C., & Khalil, R. A. (2017) Protein Kinase C as Regulator of Vascular Smooth Muscle Function and Potential Target in Vascular Disorders. *Adv Pharmacol*, 78, 203-301.

Roan, E., & Waters, C. M. (2011). What do we know about mechanical strain in lung alveoli? *Am J Physiol Lung Cell Mol Physiol*, *301*(5).

Rodríguez, H. A., Santambrosio, N., Santamaría, C. G., Muñoz-de-Toro, M., & Luque, E. H. (2010). Neonatal exposure to bisphenol A reduces the pool of primordial follicles in the rat ovary. *Reproductive Toxicology (Elmsford, N.Y.)*, *30*(4), 550–557. https://doi.org/10.1016/j.reprotox.2010.07.008.

Ross, R. (1999). Atherosclerosis–an inflammatory disease. *N Engl J Med*, *340*, 115–126.

Roszer, T. (2015). Understanding the Mysterious M2 Macrophage through Activation Markers and Effector Mechanisms. *Mediators Inflamm*, *2015*, 816460. doi: 10.1155/2015/816460

Royuela, M., Fraile, B., Arenas, M. I., & Paniagua, R. (2000). Characterization of several invertebrate muscle cell types: A comparison with vertebrate muscles. *Microsc Res Tech*, *48*(2), 107–115.

Royuela, M., Fraile, B., Picazo, M. L., & Paniagua, R. (1997). Immunocytochemical electron microscopic study and Western blot analysis of caldesmon and calponin in striated muscle of the fruit fly Drosophila melanogaster and in several muscle cell types of the earthworm Eisenia foetida. *Eur J Cell Biol*, *72*(1), 90–94.

Rutkovskiy, A., Malashicheva, A., Sullivan, G., Bogdanova, M., Kostareva, A., & Stenslokken, K. O. (2017). Valve Interstitial Cells: The Key to Understanding the Pathophysiology of Heart Valve Calcification. *J Am Heart Assoc*, *6*(9), e006339. doi: 10.1161/JAHA.117.006339.

Ryer-Powder, J. E., & Forman, H. J. (1989). Adhering lung macrophages produce superoxide demonstrated with desferal-Mn(IV. *Free Radic Biol Med*, *6*(5), 513–518.

Samaha, F. F., Ip, H. S., Morrisey, E. E., Seltzer, J., Tang, Z., Solway, J., & Parmacek, M. S. (1996). Developmental pattern of expression and genomic organization of the calponin-h1 gene. A contractile smooth muscle cell marker. *J Biol Chem*, *271*, 395–403.

Sandbo, N., & Dulin, N. (2011). Actin cytoskeleton in myofibroblast differentiation: Ultrastructure defining form and driving function. *Transl Res*, *158*(4), 181–196.

Santos-Martínez, M. J., Prina-Mello, A., Medina, C., & Radomski, M. W. (2011). Analysis of platelet function: Role of microfluidics and nanodevices. *Analyst*, *136*, 5120-5126.

Sasaki, Y., Yamamura, H., Kawakami, Y., Yamada, T., Hiratsuka, M., Kameyama, M., Ohigashi, H., Ishikawa, O., Imaoka, S., Ishiguro, S., & Takahashi, K. (2002). Expression of smooth muscle calponin in tumor vessels of human hepatocellular carcinoma and its possible association with prognosis. *Cancer*, *94*, 1777–1786.

Sathyamurthy, I., & Alex, S. (2015). Calcific aortic valve disease: Is it another face of atherosclerosis? *Indian Heart J*, *67*(5), 503–506.

Sayar, N., Karahan, G., Konu, O., Bozkurt, B., Bozdogan, O., & Yulug, I. G. (2015). Transgelin gene is frequently downregulated by promoter DNA hypermethylation in breast cancer. *Clin Epigenetics*, *7*, 104. doi: 10.1186/s13148-015-0138-5.

Schenker, T., & Trueb, B. (1998). Down-regulated proteins of mesenchymal tumor cells. *Exp Cell Res*, *239*, 161–168.

Schwarz, U. S., & Gardel, M. L. (2012). United we stand: Integrating the actin cytoskeleton and cell-matrix adhesions in cellular mechanotransduction. *J Cell Sci*, *125*(Pt 13), 3051–3060.

Senft, D., & Ronai, Z. A. (2015). UPR, autophagy, and mitochondria crosstalk underlies the ER stress response. *Trends Biochem Sci*, *40*(3), 141–148.

Seth, A., Lee, B. K., Qi, S., & Vary, C. P. (2000). Coordinate expression of novel genes during osteoblast differentiation. *J Bone Miner Res*, *15*(9), 1683–1696.

Shah, J. S., Sabouni, R., Cayton Vaught, K. C., Owen, C. M., Albertini, D. F., & Segars, J. H. (2018). Biomechanics and mechanical signaling in the ovary: A systematic review. *J Assist Reprod Genet*, *35*(7), 1135–1148.

Shapland, C., Hsuan, J. J., & Totty, N. F. (1993). Purification and properties of transgelin: A transformation and shape change sensitive actin-gelling protein. *J Cell Biol*, *121*, 1065–1073.

Sharma, A., Dagar, S., & Mylavarapu, S. V. S. (2020). Transgelin-2 and phosphoregulation of the LIC2 subunit of dynein govern mitotic spindle orientation. *J Cell Sci*, *133*(12), jcs239673. doi: 10.1242/jcs.239673.

Sharma, N., Lu, Y., Zhou, G., Liao, X., Kapil, P., Anand, P., Mahabeleshwar, G. H., Stamler, J. S., & Jain, M. K. (2012). Myeloid Kruppel-like factor 4 deficiency augments atherogenesis in ApoE-/- mice–brief report. *Arterioscler Thromb Vasc Biol*, *32*, 2836–2838.

She, Y., Li, C., Jiang, T., Lei, S., Zhou, S., Shi, H., & Chen, R. (2021). Knockdown of CNN3 Impairs Myoblast Proliferation, Differentiation, and Protein Synthesis via the mTOR Pathway. *Frontiers in Physiology*, *12*(July). https://doi.org/10.3389/fphys.2021.659272.

Sheehan, G., & Kavanagh, K. (2018). Analysis of the early cellular and humoral responses of Galleria mellonella larvae to infection by Candida albicans. *Virulence*, *9*(1), 163–172.

Shibukawa, Y., Yamazaki, N., Daimon, E., & Wada, Y. (2013). Rock-dependent calponin 3 phosphorylation regulates myoblast fusion. *Exp Cell Res*, *319*(5), 633–648. https://doi.org/10.1016/j.yexcr.2012.12.022.

Shibukawa, Y., Yamazaki, N., Kumasawa, K., Daimon, E., Tajiri, M., Okada, Y., Ikawa, M., & Wada, Y. (2010). Calponin 3 regulates actin cytoskeleton rearrangement in trophoblastic cell fusion. *Molecular Biology of the Cell*, *21*(22), 3973–3984. https://doi.org/10.1091/mbc.E10-03-0261.

Shields, J. M., Rogers-Graham, K., & Der, C. J. (2002). Loss of transgelin in breast and colon tumors and in RIE-1 cells by Ras deregulation of gene expression through Raf-independent pathways. *J Biol Chem*, *277*, 9790–9799.

Shimokawa-Kuroki, R., H, S., & Taniguchi, S. (1994). A variant actin (beta m) reduces metastasis of mouse B16 melanoma. *Int J Cancer*, *56*, 689–697.

Shirinsky, V. P., Biryukov, K. G., Hettasch, J. M., & Sellers, J. R. (1992). Inhibition of the relative movement of actin and myosin by caldesmon and calponin. *J Biol Chem*, *267*, 15886–15892.

Shishibori, T., Yamashita, K., Bandoh, J., Oyama, Y., & Kobayashi, R. (1996). Presence of Ca(2+)-sensitive and -insensitive SM22 alpha isoproteins in bovine aorta. *Biochem Biophys Res Commun*, *229*(1), 225–230.

Sider, K. L., Blaser, M. C., & Simmons, C. A. (2011). Animal models of calcific aortic valve disease. *Int J Inflam*, *2011*(364310).

Silva, M. T. (2011). Macrophage phagocytosis of neutrophils at inflammatory/infectious foci: A cooperative mechanism in the control of infection and infectious inflammation. *Journal of Leukocyte Biology*, *89*, 675-683,.

Silversides, C. K., Lionel, A. C., Costain, G., Merico, D., Migita, O., Liu, B., Yuen, T., & Rickaby, J., Thiruvahindrapuram, B., Marshall, C. R., Scherer, S. W., & Bassett, A. S. (2012). Rare copy number variations in adults with tetralogy of Fallot implicate novel risk gene pathways. *PLoS Genet*, *8:e1002843*.

Sirenko, V. V., Simonyan, A. H., Dobrzhanskaya, A. V., Shelud'ko, N. S., & Borovikov, Y. S. (2013). 40-kDa protein from thin filaments of the mussel Crenomytilus grayanus changes the conformation of F-actin during the ATPase cycle. *Biochemistry (Mosc*, *78*(3), 273–281.

Small, J. V., & Gimona, M. (1998). The cytoskeleton of the vertebrate smooth muscle cell. *Acta Physiologica Scandinavica*, *164*(4), 341–348.

Somara, S., & Bitar, K. N. (2008). Direct association of calponin with specific domains of PKC-alpha. *Am J Physiol Gastrointest Liver Physiol*, *295*, 1246–1254.

Song, S., Tan, J., Miao, Y., & Zhang, Q. (2018). Crosstalk of ER stress-mediated autophagy and ER-phagy: Involvement of UPR and the core autophagy machinery. *J Cell Physiol*, *233*(5), 3867–3874.

Sosale, N. G., Spinler, K. R., Alvey, C., & Discher, D. E. (2015). Macrophage engulfment of a cell or nanoparticle is regulated by unavoidable opsonization, a species-specific “Marker of Self” CD47, and target physical properties. *Curr Opin Immunol*, *35*, 107–112.

Sridharan, R., Cameron, A. R., Kelly, D. J., Kearney, C. J., & O’Brien, F. J. (2015). Biomaterial based modulation of macrophage polarization: A review and suggested design principles. *Materials Today*, *18*(6), 313–325.

Stables, M. J., Shah, S., Camon, E. B., Lovering, R. C., Newson, J., Bystrom, J., Farrow, S., & Gilroy, D. W. (2011). Transcriptomic analyses of murine resolution-phase macrophages. *Blood*, *118*, 192-208,.

Stalker, T. J., Traxler, E. A., Wu, J., Wannemacher, K. M., Cermignano, S. L., Voronov, R., Diamond, S. L., & Brass, L. F. (2013). Hierarchical organization in the hemostatic response and its relationship to the platelet-signaling network. *Blood*, *121*, 1875-1885,.

Stewart, B. F., Siscovick, D., Lind, B. K., Gardin, J. M., Gottdiener, J. S., & Smith, V. E. (1997). Clinical factors associated with calcific aortic valve disease. *Cardiovascular Health Study. J Am Coll Cardiol*, *29*(3), 630–634.

Stork, N. E. (2018). How Many Species of Insects and Other Terrestrial Arthropods Are There on Earth? *Annu Rev Entomol*, *63*, 31–45.

Stow, J. L., & Condon, N. D. (2016). The cell surface environment for pathogen recognition and entry. *Clin Transl Immunology*, *5*(4), e71. doi: 10.1038/cti.2016.15.

St-Pierre, J., Moreau, F., Cornick, S., Quach, J., Begum, S., Aracely Fernandez, L., Gorman, H., & Chadee, K. (2017). The macrophage cytoskeleton acts as a contact sensor upon interaction with Entamoeba histolytica to trigger IL-1beta secretion. *PLoS Pathog*, *13*(8), e1006592. doi: 10.1371/journal.ppat.1006592.

Strasser, P., Gimona, M., Moessler, H., Herzog, M., & Small, J. V. (1993). Mammalian calponin. Identification and expression of genetic variants. *FEBS Letters*, *330*(1), 13–18. https://doi.org/10.1016/0014-5793(93)80909-e.

Strik, C., Wever, K. E., Stommel, M. W. J., Goor, H. V., & Ten Broek, R. P. G. (2019). Adhesion reformation and the limited translational value of experiments with adhesion barriers: A systematic review and meta-analysis of animal models. *Sci Rep*, *9*(1). https://doi.org/10.1038/s41598-019-52457-2.

Sugenoya, Y., Yoshimura, A., Yamamura, H., Inui, K., Morita, H., Yamabe, H., Ueki, N., Ideura, T., & Takahashi, K. (2002). Smooth-muscle calponin in mesangial cells: Regulation of expression and a role in suppressing glomerulonephritis. *J Am Soc Nephrol*, *13*, 322–331.

Suleiman, H. Y., Roth, R., Jain, S., Heuser, J. E., Shaw, A. S., & Miner, J. H. (2017). Injury-induced actin cytoskeleton reorganization in podocytes revealed by super-resolution microscopy. *JCI Insight*, *2*(16), e94137, 94137. https://doi.org/10.1172/jci.insight.94137.

Sun, C., Yang, X., Wang, T., Cheng, M., & Han, Y. (2021). Ovarian Biomechanics: From Health to Disease. *Frontiers in Oncology*, *11*, 744257. https://doi.org/10.3389/fonc.2021.744257.

Szekanecz, Z., & Koch, A. E. (2007). Macrophages and their products in rheumatoid arthritis. *Current Opinion in Rheumatology*, *19*(3), 289–295. https://doi.org/10.1097/BOR.0b013e32805e87ae.

Szymanski, P. T., & Tao, T. (1997). Localization of protein regions involved in the interaction between calponin and myosin. *J Biol Chem*, *272*, 11142–11146.

Takahashi, K., Abe, M., Hiwada, K., & Kokubu, T. (1988). A novel troponin T-like protein (calponin) in vascular smooth muscle: Interaction with tropomyosin paracrystals. *J Hypertens Suppl*, *6*, 40–43.

Takahashi, K., Hiwada, K., & Kokubu, T. (1986). Isolation and characterization of a 34,000-dalton calmodulin- and F-actin-binding protein from chicken gizzard smooth muscle. *Biochem Biophys Res Commun*, *141*(1), 20–26.

Takahashi, K., Hiwada, K., & Kokubu, T. (1988). Vascular smooth muscle calponin. A novel troponin T-like protein. *Hypertension*, *11*(6 Pt 2), 620–626. https://doi.org/10.1161/01.hyp.11.6.620

Takahashi, K., Yoshimoto, R., Fuchibe, K., Fujishige, A., Mitsui-Saito, M., Hori, M., Ozaki, H., Yamamura, H., Awata, N., Taniguchi, S. ichiro, Katsuki, M., Tsuchiya, T., & Karaki, H. (2000). Regulation of shortening velocity by calponin in intact contracting smooth muscles. *Biochemical and Biophysical Research Communications*, *279*(1), 150–157. https://doi.org/10.1006/bbrc.2000.3909.

Takemoto, K., Ishihara, S., Mizutani, T., Kawabata, K., & Haga, H. (2015). Compressive stress induces dephosphorylation of the myosin regulatory light chain via RhoA phosphorylation by the adenylyl cyclase/protein kinase a signaling pathway. *PLoS ONE*, *10*(3), 1–17. https://doi.org/10.1371/journal.pone.0117937.

Takeoka, M., Ehara, T., Sagara, J., Hashimoto, S., & Taniguchi, S. (2002). Calponin h1 induced a flattened morphology and suppressed the growth of human fibrosarcoma HT1080 cells. *Eur J Cancer*, *38*, 436–442.

Takeuchi, K., Takahashi, K., Abe, M., Nishida, W., Hiwada, K., Nabeya, T., & Maruyama, K. (1991). Co-localization of immunoreactive forms of calponin with actin cytoskeleton in platelets, fibroblasts, and vascular smooth muscle. *Journal of Biochemistry*, *109*, 311-316,.

Tanaka, Y., Nakayamada, S., & Okada, Y. (2005). Osteoblasts and osteoclasts in bone remodeling and inflammation. *Current Drug Targets Inflammation and Allergy*, *4*, 325-328,.

Tang, D. C., Kang, H. M., Jin, J. P., Fraser, E. D., & Walsh, M. P. (1996). Structure-function relations of smooth muscle calponin: The critical role of serine 175. *Journal of Biological Chemistry*, *271*(15), 8605–8611. https://doi.org/10.1074/jbc.271.15.8605.

Tang, J., Hu, G., Hanai, J. I., Yadlapalli, G., Lin, Y., Zhang, B., Galloway, J., Bahary, N., Sinha, S., Thisse, B., Thisse, C., Jin, J. P., Zon, L. I., & Sukhatme, V. P. (2006). A critical role for calponin 2 in vascular development. *Journal of Biological Chemistry*, *281*(10), 6664–6672. https://doi.org/10.1074/jbc.M506991200.

Tangirala, R. K., Rubin, E. M., & Palinski, W. (1995). Quantitation of atherosclerosis in murine models: Correlation between lesions in the aortic origin and in the entire aorta, and differences in the extent of lesions between sexes in LDL receptor-deficient and apolipoprotein E-deficient mice. *J Lipid Res*, *36*, 2320–2328.

Taniguchi, S. (2005). Suppression of cancer phenotypes through a multifunctional actin-binding protein, calponin, that attacks cancer cells and simultaneously protects the host from invasion. *Cancer Sci*, *96*, 738-746.

Tas, S. W., Quartier, P., Botto, M., & Fossati-Jimack, L. (2006). Macrophages from patients with SLE and rheumatoid arthritis have defective adhesion in vitro, while only SLE macrophages have impaired uptake of apoptotic cells. *Annals of the Rheumatic Diseases*, *65*, 216-221.

Teixeira, A. I., Ilkhanizadeh, S., Wigenius, J. A., Duckworth, J. K., Inganas, O., & Hermanson, O. (2009). The promotion of neuronal maturation on soft substrates. *Biomaterials*, *30*, 4567–4572.

ten Broek, R. P., Bakkum, E. A., Laarhoven, C. J. H. M., & van Goor, H. (2016). Epidemiology and Prevention of Postsurgical Adhesions Revisited. *Annals of Surgery*, *263*(1), 12–19.

Thomas, C. M., & Smart, E. J. (2007). Gender as a regulator of atherosclerosis in murine models. *Curr Drug Targets*, *8*, 1172–1180.

Thompson, O., Moghraby, J. S., Ayscough, K. R., & Winder, S. J. (2012). Depletion of the actin bundling protein SM22/transgelin increases actin dynamics and enhances the tumourigenic phenotypes of cells. *BMC Cell Biol*, *13*(1), doi: 10.1186/1471-2121-13-1.

Towler, D. A., Shao, J. S., Cheng, S. L., Pingsterhaus, J. M., & Loewy, A. P. (2006). Osteogenic regulation of vascular calcification. *Ann N Y Acad Sci*, *1068*, 327–333.

Trabelsi-Terzidis, H., Fattoum, A., Represa, A., Dessi, F., Ben-Ari, Y., & Terrossian, E. (1995). Expression of an acidic isoform of calponin in rat brain: Western blots on one- or two-dimensional gels and immunolocalizatian in cultured cells. *Biochemical Journal*, *306*(1), 211–215. https://doi.org/10.1042/bj3060211.

Tschumperlin, D. J., Boudreault, F., & Liu, F. (2010). Recent advances and new opportunities in lung mechanobiology. *J Biomech*, *43*(1), 99–107.

Tulandi, T., Agdi, M., Zarei, A., Miner, L., & Sikirica, V. (2009). Adhesion development and morbidity after repeat cesarean delivery. *Am J Obstet Gynecol. Jul*, *201*(1), 1–6. https://doi.org/10.1016/j.ajog.2009.04.039.

Turner, C. E. (2000). Paxillin and focal adhesion signalling. *Nat Cell Biol*, *2*, 231–236.

Tuxhorn, J. A., Ayala, G. E., Smith, M. J., Smith, V. C., Dang, T. D., & Rowley, D. R. (2002). Reactive stroma in human prostate cancer: Induction of myofibroblast phenotype and extracellular matrix remodeling. *Clin Cancer Res*, *8*, 2912–2923.

Tzima, E. (2006). Role of small GTPases in endothelial cytoskeletal dynamics and the shear stress response. *Circ Res*, *98*, 176–185.

Ulmer, B., Hagenlocher, C., & Schmalholz, S. (2013). Calponin 2 acts as an effector of noncanonical Wnt-mediated cell polarization during neural crest cell migration. *Cell Rep*, *3*(3), 615–621.

Uzquiano, M. C., Prieto, V. G., Nash, J. W., Ivan, D. S., Gong, Y., Lazar, A. J., & Diwan, A. H. (2008). Metastatic basal cell carcinoma exhibits reduced actin expression. *Mod Pathol*, *21*, 540–543.

van Gils, J. M., Derby, M. C., Fernandes, L. R., Ramkhelawon, B., Ray, T. D., Rayner, K. J., Parathath, S., Distel, E., Feig, J. L., Alvarez-Leite, J. I., Rayner, A. J., McDonald, T. O., O'Brien, K. D., Stuart, L. M., Fisher, E. A., Lacy-Hulbert, A., & Moore, K. J. (2012). The neuroimmune guidance cue netrin-1 promotes atherosclerosis by inhibiting the emigration of macrophages from plaques. *Nature Immunology*, *13*(2), 136–143. https://doi.org/10.1038/ni.2205.

van Goor, H. (2007). Consequences and complications of peritoneal adhesions. *Colorectal Disease: The Official Journal of the Association of Coloproctology of Great Britain and Ireland*, *9 Suppl 2*, 25–34. https://doi.org/10.1111/j.1463-1318.2007.01358.x.

van Vliet, A. R., Giordano, F., Gerlo, S., Segura, I., Van Eygen, S., Molenberghs, G., Rocha, S., Houcine, A., Derua, R., Verfaillie, T., Vangindertael, J., De Keersmaecker, H., Waelkens, E., Tavernier, J., Hofkens, J., Annaert, W., Carmeliet, P., Samali, A., Mizuno, H., & Agostinis, P. (2017). The ER Stress Sensor PERK Coordinates ER-Plasma Membrane Contact Site Formation through Interaction with Filamin-A and F-Actin Remodeling. *Molecular Cell*, *65*(5), 885-899.e6. https://doi.org/10.1016/j.molcel.2017.01.020.

Vaz, B., El Mansouri, F., Liu, X., & Taketo, T. (2020). Premature ovarian insufficiency in the XO female mouse on the C57BL/6J genetic background. *Molecular Human Reproduction*, *26*(9), 678–688. https://doi.org/10.1093/molehr/gaaa049.

Verone, A. R., Duncan, K., & Godoy, A. (2013). Androgen-responsive serum response factor target genes regulate prostate cancer cell migration. *Carcinogenesis*, *34*(8), 1737–1746.

Vetter, S. W., & Leclerc, E. (2003). Novel aspects of calmodulin target recognition and activation. *European Journal of Biochemistry*, *270*(3), 404–414. https://doi.org/10.1046/j.1432-1033.2003.03414.x.

Wage, J., Lerebours, A., Hardege, J. D., & Rotchell, J. M. (2016). Exposure to low pH induces molecular level changes in the marine worm, Platynereis dumerilii. *Ecotoxicol Environ Saf*, *124*, 105–110.

Waite, A. L., Schaner, P., Hu, C., Richards, N., Balci-Peynircioglu, B., Hong, A., Fox, M., & Gumucio, D. L. (2009). Pyrin and ASC co-localize to cellular sites that are rich in polymerizing actin. *Exp Biol Med (Maywood, 234*(1), 40–52.

Walker, G. A., Masters, K. S., Shah, D. N., Anseth, K. S., & Leinwand, L. A. (2004). Valvular myofibroblast activation by transforming growth factor-beta: Implications for pathological extracellular matrix remodeling in heart valve disease. *Circ Res*, *95*(3), 253–260.

Walker, J. L., Fournier, A. K., & Assoian, R. K. (2005). Regulation of growth factor signaling and cell cycle progression by cell adhesion and adhesion-dependent changes in cellular tension. *Cytokine Growth Factor Rev*, *16*, 395–405.

Walsh, M. P. (1991). The Ayerst Award Lecture 1990. Calcium-dependent mechanisms of regulation of smooth muscle contraction. *Biochem Cell Biol*, *69*, 771–800.

Walsh, M. P., & Cole, W. C. (2013). The role of actin filament dynamics in the myogenic response of cerebral resistance arteries. *Journal of Cerebral Blood Flow and Metabolism: Official Journal of the International Society of Cerebral Blood Flow and Metabolism*, *33*(1), 1–12.

Wang, J., & Kubes, P. (2016). A Reservoir of Mature Cavity Macrophages that Can Rapidly Invade Visceral Organs to Affect Tissue Repair. *Cell*, *165*(3), 668–678. https://doi.org/10.1016/j.cell.2016.03.009.

Wang, N., Tytell, J. D., & Ingber, D. E. (2009). Mechanotransduction at a distance: Mechanically coupling the extracellular matrix with the nucleus. *Nat Rev Mol Cell Biol*, *10*, 75–82.

Wang, P., & Gusev, N. B. (1996). Interaction of smooth muscle calponin and desmin. *FEBS Letters*, *392*(3), 255–258. https://doi.org/10.1016/0014-5793(96)00824-1.

Warnes, G. (2014). Measurement of autophagy by flow cytometry. *Curr Protoc Cytom*, *68*(9), 45 1-10.

Weigert, A., Johann, A. M., Knethen, A., Schmidt, H., Geisslinger, G., & Brune, B. (2006). Apoptotic cells promote macrophage survival by releasing the antiapoptotic mediator sphingosine-1-phosphate. *Blood*, *108*, 1635-1642.

Weissgerber, T. L., Craici, I. M., Wagner, S. J., Grande, J. P., & Garovic, V. D. (2014). Advances in the pathophysiology of preeclampsia and related podocyte injury. *Kidney Int*, *86*, 445.

Wesley, R. B., Meng, X., Godin, D., & Galis, Z. S. (1998). Extracellular matrix modulates macrophage functions characteristic to atheroma: Collagen type I enhances acquisition of resident macrophage traits by human peripheral blood monocytes in vitro. *Arteriosclerosis, Thrombosis, and Vascular Biology*, *18*(3), 432–440. https://doi.org/10.1161/01.atv.18.3.432.

Whang, S. H., Astudillo, J. A., & Sporn, E. (2011). In search of the best peritoneal adhesion model: Comparison of different techniques in a rat model. *J Surg Res. May*, *167*(2), 245–250. https://doi.org/10.1016/j.jss.2009.06.020.

White, J., Lancelot, M., Sarnaik, S., & Hines, P. (2014). Increased erythrocyte adhesion to VCAM-1 during pulsatile flow: Application of a microfluidic flow adhesion bioassay. *Clinical Hemorheology and Microcirculation,* 60(2), 201-13. doi: 10.3233/CH-141847.

Wills, F. L., McCubbin, W. D., & Kay, C. M. (1993). Characterization of the Smooth Muscle Calponin and Calmodulin Complex. *Biochemistry, 32*(9), 2321–2328. https://doi.org/10.1021/bi00060a025.

Winder, S. J., Allen, B. G., Fraser, E. D., Kang, H. M., Kargacin, G. J., & Walsh, M. P. (1993). Calponin phosphorylation in vitro and in intact muscle. *Biochemical Journal, 296*(3), 827–836. https://doi.org/10.1042/bj2960827.

Winder, S. J., Allen, B. G., O, C.-C., & Walsh, M. P. (1998). Regulation of smooth muscle actin-myosin interaction and force by calponin. *Acta Physiol Scand, 164*, 415–426.

Winder, S. J., Jess, T., & Ayscough, K. R. (2003). SCP1 encodes an actin-bundling protein in yeast. *Biochem J, 375*(Pt 2), 287–295.

Winder, S. J., & Walsh, M. P. (1990). Smooth muscle calponin. Inhibition of actomysosin MgATPase and regulation by phosphorylation. *Journal of Biological Chemistry, 265*(17), 10148–10155. https://doi.org/10.1016/s0021-9258(19)38792-7.

Winder, S. J., & Walsh, M. P. (1993). Calponin: Thin filament-linked regulation of smooth muscle contraction. *Cellular Signalling, 5*(6), 677–686. https://doi.org/10.1016/0898-6568(93)90029-l

Winder, S. J., Walsh, M. P., Vasulka, C., & Johnson, J. D. (1993). Calponin-Calmodulin Interaction: Properties and Effects on Smooth and Skeletal Muscle Actin Binding and Actomyosin ATPases. *Biochemistry, 32*(48), 13327–13333. https://doi.org/10.1021/bi00211a046.

Wipff, P. J., Rifkin, D. B., Meister, J. J., & Hinz, B. (2007). Myofibroblast contraction activates latent TGF-beta1 from the extracellular matrix. *J Cell Biol, 179*(6), 1311–1323.

Wirrig, E. E., & Yutzey, K. E. (2014). Conserved transcriptional regulatory mechanisms in aortic valve development and disease. *Arterioscler Thromb Vasc Biol, 34*(4), 737–741.

Wu, K.-C., & Jin, J.-P. (2008). Calponin in non-muscle cells. *Cell Biochemistry and Biophysics, 52*(3), 139–148. https://doi.org/10.1007/s12013-008-9031-6.

Xu, J., & Gridley, T. (2013). Notch2 is required in somatic cells for breakdown of ovarian germ-cell nests and formation of primordial follicles. *BMC Biology, 11*, 13. https://doi.org/10.1186/1741-7007-11-13.

Yamamura, H., Hashio, M., Noguchi, M., Sugenoya, Y., Osakada, M., Hirano, N., Sasaki, Y., Yoden, T., Awata, N., Araki, N., Tatsuta, M., Miyatake, S. I., & Takahashi, K. (2001). Identification of the transcriptional regulatory sequences of human calponin promoter and their use in targeting a conditionally replicating herpes vector to malignant human soft tissue and bone tumors. *Cancer Res, 61*, 3969–3977.

Yamamura, Y., Shelden, E, Fox, D.A. (2000). Interactions between T cells and synovial fibroblasts. *Modern rheumatology/the Japan Rheumatism Ass.ociation* 10: 16-18.

Yamashiro, S., Gimona, M., & Ono, S. (2007). UNC-87, a calponin-related protein in C. elegans, antagonizes ADF/cofilin-mediated actin filament dynamics. *J Cell Sci, 120*(Pt 17), 3022–3033.

Yan, C., Wang, P., DeMayo, J., DeMayo, F. J., Elvin, J. A., Carino, C., Prasad, S. V., Skinner, S. S., Dunbar, B. S., Dube, J. L., Celeste, A. J., & Matzuk, M. M. (2001). Synergistic roles of bone morphogenetic protein 15 and growth differentiation factor 9 in ovarian function. *Molecular Endocrinology (Baltimore, Md.), 15*(6), 854–866. https://doi.org/10.1210/mend.15.6.0662.

Yang, C., Zhu, S., Feng, W., & Chen, X. (2021). Calponin 3 suppresses proliferation, migration and invasion of non-small cell lung cancer cells. *Oncology Letters*, *22*(2), 634. https://doi.org/10.3892/ol.2021.12895.

Yang, W., Zheng, Y. Z., Jones, M. K., & McManus, D. P. (1999). Molecular characterization of a calponin-like protein from Schistosoma japonicum. *Molecular and Biochemical Parasitology*, *98*(2), 225–237. https://doi.org/10.1016/s0166-6851(98)00171-6.

Yang, X., Meng, X., Su, X., Mauchley, D. C., Ao, L., & Cleveland, J. C., Jr. (2009). Bone morphogenic protein 2 induces Runx2 and osteopontin expression in human aortic valve interstitial cells: Role of Smad1 and extracellular signal-regulated kinase 1/2. *J Thorac Cardiovasc Surg*, *138*(4), 1008–1015.

Yang, Y. L., Lee, M. G., & Lee, C. C. (2018). Pentoxifylline decreases post-operative intra-abdominal adhesion formation in an animal model. *PeerJ*, *6:e5434*. https://doi.org/10.7717/peerj.5434.

Yang, Z., Chang, Y. J., & Miyamoto, H. (2007). Transgelin functions as a suppressor via inhibition of ARA54-enhanced androgen receptor transactivation and prostate cancer cell growth. *Mol Endocrinol*, *21*(2), 343–358.

Yap, A. S., Duszyc, K., & Viasnoff, V. (2018). Mechanosensing and Mechanotransduction at Cell-Cell Junctions. *Cold Spring Harbor Perspectives in Biology*, *10*(8), a028761. https://doi.org/10.1101/cshperspect.a028761.

Ye, L., Kynaston, H. G., & Jiang, W. G. (2007). Bone metastasis in prostate cancer: Molecular and cellular mechanisms. *J Mol Med*, *20*, 103–111.

Yin, L. M., Xu, Y. D., Peng, L. L., Duan, T. T., Liu, J. Y., Xu, Z., Wang, W. Q., Guan, N., Han, X. J., Li, H. Y., Pang, Y., Wang, Y., Chen, Z., Zhu, W., Deng, L., Wu, Y. L., Ge, G. B., Huang, S., Ulloa, L., & Yang, Y. Q. (2018). Transgelin-2 as a therapeutic target for asthmatic pulmonary resistance. *Sci Transl Med*, *10*(427), eaam8604. doi: 10.1126/scitranslmed.aam8604.

Yip, C. Y., Chen, J. H., Zhao, R., & Simmons, C. A. (2009). Calcification by valve interstitial cells is regulated by the stiffness of the extracellular matrix. *Arterioscler Thromb Vasc Biol*, *29*(6), 936–942.

Yu, Y. R., O'Koren, E. G., Hotten, D. F., Kan, M. J., Kopin, D., Nelson, E. R., Que, L., & Gunn, M. D. (2016). A Protocol for the Comprehensive Flow Cytometric Analysis of Immune Cells in Normal and Inflamed Murine Non-Lymphoid Tissues. *PLoS One*, *11*(3), e0150606. doi: 10.1371/journal.pone.0150606.

Yutzey, K. E., Demer, L. L., Body, S. C., Huggins, G. S., Towler, D. A., & Giachelli, C. M. (2014). Calcific aortic valve disease: A consensus summary from the Alliance of Investigators on Calcific Aortic Valve Disease. *Arterioscler Thromb Vasc Biol*, *34*(11), 2387–2393.

Yvan-Charvet, L., Pagler, T., Gautier, E. L., Avagyan, S., Siry, R. L., Han, S., Welch, C. L., Wang, N., Randolph, G. J., Snoeck, H. W., & Tall, A. R. (2010). ATP-binding cassette transporters and HDL suppress hematopoietic stem cell proliferation. *Science*, *328*, 1689–1693.

Zeidan, A., Nordstrom, I., & Albinsson, S. (2003). Stretch-induced contractile differentiation of vascular smooth muscle: Sensitivity to actin polymerization inhibitors. *Am J Physiol Cell Physiol*, *284*, 1387–1396.

Zeidan, A., Nordstrom, I., & Dreja, K. (2000). Stretch-dependent modulation of contractility and growth in smooth muscle of rat portal vein. *Circ Res*, *87*, 228–234.

Zerbinatti, C. V., & Gore, R. W. (2003). Uptake of modified low-density lipoproteins alters actin distribution and locomotor forces in macrophages. *Am J Physiol Cell Physiol*, *284*, 555–561.

Zhang, X., Meng, H., & Wang, M. M. (2013). Collagen represses canonical Notch signaling and binds to Notch ectodomain. *Int J Biochem Cell Biol*, *45*(7), 1274–1280.

Zhang, X., & Mosser, D. M. (2008). Macrophage activation by endogenous danger signals. *J Pathol*, *214*(2), 161–178.

Zhang, Y., Ye, Y., & Shen, D. (2010). Identification of transgelin-2 as a biomarker of colorectal cancer by laser capture microdissection and quantitative proteome analysis. *Cancer Sci*, *101*(2), 523–529.

Zhong, L., He, X., & Si, X. (2019). SM22alpha (Smooth Muscle 22alpha) Prevents Aortic Aneurysm Formation by Inhibiting Smooth Muscle Cell Phenotypic Switching Through Suppressing Reactive Oxygen Species/NF-kappaB (Nuclear Factor-kappaB. *Arterioscler Thromb Vasc Biol*, *39*(1), e10-e25. doi: 10.1161/ATVBAHA.118.311917.

Zhou, L. Y., Jin, C. X., Wang, W. X., Song, L., Shin, J. B., Du, T. T., & Wu, H. (2023). Differential regulation of hair cell actin cytoskeleton mediated by SRF and MRTFB. *Elife,* 12, e90155. doi: 10.7554/eLife.90155.

Zhu, Q., Emanuele, N. V., & Thiel, D. H. (2004). Calponin is expressed by Sertoli cells within rat testes and is associated with actin-enriched cytoskeleton. *Cell Tissue Res*, *316*, 243–253.

About the Authors

Jian-Ping (J.-P.) Jin, MD, MSc, PhD, is Professor of Physiology and Biophysics at University of Illinois at Chicago College of Medicine. Before joining UIC, He was the William D. Traitel Endowed Professor and Chairman of the Department of Physiology at Wayne State University School of Medicine from 2009 to 2020 and Chief of the Section of Molecular Cardiology at Evanston Northwestern Healthcare and Professor of Medicine at Northwestern University Feinberg School of Medicine from 2004 to 2009. His prior academic appointments include Associate Professor of Physiology and Biophysics at Case Western Reserve University from 1996 to 2003 and Assistant Professor of Biochemistry and Molecular Biology at the University of Calgary in Canada from 1993 to 1996. He completed his medical education and cardiology residency in Xi'an, China, and received his PhD in Biology from University of Iowa in 1989. Dr. Jin is a leading researcher in the field of muscle contractility and cell motility. His research interest is in the gene regulation, structure-function relationship and pathophysiological adaptation of contractile and cytoskeleton proteins, focusing on the actin thin filament regulatory proteins troponin and calponin. He has published more than 190 peer-reviewed research papers and 28 review articles. His research discoveries have led to translational applications in the diagnosis of myocardial ischemia and the treatments of heart failure and cancer metastasis as described in multiple United States patents. Dr. Jin was inducted into the Wayne State University Academy of Scholars in 2016 as a Life Member that is the highest recognition bestowed upon fellow faculty members by their colleagues.

Qi-Quan Huang, MD, MSc, MA, is Research Professor of Medicine in the Division of Rheumatology at Northwestern University Feinberg School of Medicine. Dr. Huang graduated from the Shaanxi Chinese Medical University in China and completed a graduate research training in Microbiology & Immunology in 1982. After working as Lecturer of Medical Microbiology & Immunology from 1982 to 1989, she earned a second master's degree in Biological Sciences at University of Texas at Austin in 1991. After a postdoctoral training in Molecular Biology at the University of Alberta in Canada, she worked as Senior Research Associate in the Department of Medical Biochemistry at University of Calgary in Canada and in the Department of Physiology & Biophysics at Case Western Reserve University from 1993 to 2003. In 2004, she was appointed to Research Assistant Professor in the Department of Medicine at Northwestern University Feinberg School of Medicine, and promoted to Research Associate Professor in 2011 and Research Professor in 2019. Dr. Huang has published extensively in biomedical research, especially in inflammatory arthritis studies, including high impact articles in the journals of *Nature, Science, Nature Communication*, *Science Advance*, *Blood* and *Journal of Immunology.*

Index

H

I

K

L

M